P9-DGL-267

21st-century electronic projects for a new age

Delton T. Horn

FIRST EDITION
FIRST PRINTING

Library of Congress Cataloging-in-Publication Data

Horn, Delton T.
21st-century electronic projects / by Delton T. Horn.
p. cm.
Includes index.
ISBN 0-8306-3806-7 (H) ISBN 0-8306-3805-9 (P)
1. Electronic apparatus and appliances—Design and construction--Amateurs' manuals. 2. Testing—Equipment and supplies—Design and construction—Amateurs' manuals. I. Title.
TK9965.H62 1992
621.381—dc20 92-7489
CIP

TAB Books offers software for sale. For information and a catalog, please contact TAB Software Department, Blue Ridge Summit, PA 17294-0850.

Acquisitions Editor: Roland S. Phelps
Book Editor: Laura J. Bader
Director of Production: Katherine G. Brown
Book Design: Jaclyn J. Boone EL2

Contents

Preface *ix*

❖ 1 **Introduction to New Age** *1*
New Age and science *1*
New Age beliefs and concepts *3*
New Age and quantum physics *9*

❖ 2 **Meditation and hypnosis aids** *13*
Benefits of meditation *13*
Hypnosis *14*
Simple audio hypnotizer *17*
Improved audio hypnotizer *21*
Visual hypnotic aid *22*
Deluxe audio/visual hypnotic aid *27*
Alpha glasses *29*

❖ 3 **Biofeedback monitors** *35*
Biofeedback theory *35*
Simple skin resistance biofeedback monitor *38*
Computerized biofeedback monitor *41*
Biofeedback temperature monitor *43*
Capacitive biofeedback monitor *47*
Alpha-wave biofeedback monitor *51*
Some final words about biofeedback *56*

❖ 4 **ESP testers** *57*
Types of ESP *57*
Testing ESP *59*
Two-choice ESP tester *60*
Electronic dice *62*

Random-number generator *66*
Manual telepathy tester *69*
Automated ESP tester *74*

❖ 5 **Air ionizer** ***79***
Ions *79*
The effects of air ionization on human beings *80*
Cleaning the air *82*
Negative-ion generators *83*
Negative-ion generator project *85*
Cautions and safety considerations *88*

❖ 6 **Biorhythms** ***91***
What are biorhythms? *91*
Rhythms in life *94*
Flaws in the reasoning *97*
Some historical background *99*
Biorhythm calculation program *102*
Biorhythm clock *108*

❖ 7 **Kirlian photography** ***119***
Nonstandard photography *119*
Kirlian photography *120*
Auras *121*
Possible applications *123*
The phantom leaf effect *124*
More on diagnostic aura research *125*
The skeptics *127*
The experimental circuit *128*
Making the Kirlian photograph *132*
Safety precautions *135*

❖ 8 **Detecting UFOs** ***137***
Beliefs about UFOs *137*
Magnetic effects of UFOs *143*
Magnetic field detector project 1 *144*
Magnetic field detector project 2 *149*

❖ 9 **Crystals** ***155***
What is a crystal? *155*
Crystals in electronics *160*
New Age crystals *164*
But is it science? *165*
Psionic generators *172*

❖ 10 **The quantum connection** ***173***
New and old physics *174*
The peculiar nature of quantum particles *175*
The double-slit experiment *176*
Quantum movement *180*
The indeterminacy principle *181*

Index ***185***

Projects

1	Simple audio hypnotizer	*17*
2	Improved audio hypnotizer	*21*
3	Visual hypnotic aid	*22*
4	Dual LED visual hypnotic aid	*25*
5	Deluxe audible/visual hypnotic aid	*27*
6	Alpha glasses	*29*
7	Simple skin resistance biofeedback monitor	*38*
8	Computerized biofeedback monitor	*41*
9	Biofeedback temperature monitor	*43*
10	Capacitive biofeedback monitor	*47*
11	Alpha-wave biofeedback monitor	*51*
12	Two-choice ESP tester	*60*
13	Electronic dice	*62*
14	Random number generator	*66*
15	Manual telepathy tester	*69*
16	Automated ESP tester	*74*
17	Negative ion generators	*83*
18	Biorhythm calculation program	*102*
19	Biorhythm clock	*108*
20	Kirlian photography experimental circuit	*128*
21	Magnetic field detector 1	*144*
22	Magnetic field detector 2	*149*

Preface

TWENTY-FIRST CENTURY ELECTRONIC PROJECTS FOR A NEW AGE, in many ways, is a very different book than most books of electronic projects. All of the projects in this book are connected in some way to controversial New Age concepts.

A small but growing cultural phenomenon today is the "New Age" movement. Many people have heard the term, but most don't know what it means or fully understand its concepts.

While much of the New Age is purely of the mystic or spiritual realm, many New Age ideas resemble scientific hypotheses. For various reasons, most of these New Age hypotheses have never been tested scientifically. Most traditional scientists reject New Age ideas as not worth investigating. And it seems that most New Agers are willing to believe these ideas for purely philosophical reasons. They either don't care about scientific evidence or they are content to believe casual anecdotal evidence, which can never be truly conclusive. The few laboratory experiments that have been conducted into these areas are often incomplete and lack the proper scientific controls needed to ensure the validity of the resulting data.

Much has been written and published in recent years on New Age topics. Very little of this material has any real scientific usefulness or validity. Invariably, it seems, books about New Age concepts are written by "true believers." The authors are almost always trying to win converts to their personal beliefs, and they sometimes consider slight or questionable evidence as conclusive proof.

All too often, authors who believe in New Age ideas are borderline fanatics, accepting any marginally related anecdote as inarguable proof. They seem to feel that if they prove something is possible, nothing else is needed. If it is possible, they claim, then it must be true.

The debunking authors, arguing against New Age concepts, are sometimes just as bad. They accept current scientific theory as established and almost sacred fact. But the history of science is filled with long-held theories that were eventually disproved. The debunkers also seem to feel that if they show some way that an apparent phenomenon can be faked, they have done their job. If it is possible to fake, or if it has ever been faked, the debunkers claim, then all similar incidents must be faked. Curiously, many well-respected scientists have allowed themselves to slip into this kind of antiscientific trap.

In this book, we will look at a number of New Age ideas that can be tested and demonstrated (or perhaps just simulated) with electronic circuits. In discussing the New Age concepts behind these unusual projects, I have made every effort to be as fair and nondogmatic as possible. Of course, it is impossible for any human being to be completely unbiased. We all have our own opinions, which inevitably color the way we see things. I'm sure some of my personal beliefs on the subjects covered here will show through. But this book is not about what I believe. (Who cares what I believe?) I am not out to convince others to believe what I do about these matters.

My general theme and approach can be stated as follows: Just because something has not been proved does not mean it has been disproved, and just because something has not been disproved does not mean it has been proved. Often, the true scientific attitude is, ''No one knows.''

The recent peculiar and puzzling discoveries of quantum physics are slowly teaching us the folly of scientific arrogance. Many of science's old, most cherished beliefs have been upset or destroyed by quantum physics. In some cases, quantum physicists and New Agers sound like they might be talking about the same things using different terms. Relevant issues from quantum physics are also covered in this volume.

The order in which the various New Age subjects are presented here is somewhat arbitrary. I have tried to organize the topics from the more plausible to the more far-fetched, but of course, such categorization is purely a matter of opinion.

Simply put, this book is about some fascinating ideas that may or may not be true. I hope you find the subject matter as stimulating and interesting as I do. If nothing else, you should find some unusual electronic projects here. Most of these circuits would make excellent science fair projects and many of the basic circuits can be modified for other, more practical applications.

1 Introduction to New Age

MOST PEOPLE TODAY HAVE HEARD THE PHRASE "NEW AGE," BUT the majority do not really know what it means. According to a number of surveys, the general public's attitude toward New Age is vague distrust and dislike.

Just what is New Age and what does it have to do with electronics? I will attempt to answer these questions in this chapter. Real technophiles might be disappointed that there is so little electronics in this chapter, but I feel it is necessary to cover the background of the ideas behind the projects presented in the later chapters of this book. Consider this discussion as a form of theory.

New Age and science

"New Age" is a loosely applied umbrella title covering a number of more or less independent movements and ideas. Someone involved in one aspect of New Age might have no interest in (and even contempt for) aspects of New Age that others consider as absolutely fundamental and essential. It is not really appropriate to group all New Agers together as if they were a homogeneous, organized group.

For that matter, the same is true of science as well, isn't it? After all, a physicist is not a biologist, who is not a chemist. Even particular branches of science have different specialties. One scientist might have little or no interest in another's area of specialization. Some scientists even feel that certain specializations are worthless or even ridiculous. The New Age has similar diversity.

But there are certain characteristics that seem to be common to most, if not all, New Age ideas. Many New Age concepts have a strong spiritual, although usually not religious (in the traditional sense), element. An underlying assumption is that there is more to reality than we can grasp directly with our physical senses. New Agers do not believe in the strictly mechanistic and deterministic "clockwork" universe of classical physics. Before writing them off as a bunch of superstitious nuts, remember that quantum physics has pretty much overthrown the purely deterministic, mechanical universe view. There does seem to be more to reality than meets the eye (or our other senses). Of course, quantum physics deals with subatomic particles, not souls, but as we shall see, some striking similarities exist between New Age and quantum physics, despite the very pronounced differences in terminology and focus.

New Age ideas are generally not proven or accepted by traditional science. Many of us with scientific backgrounds tend to automatically discredit such ideas. But "unproved" does not mean "disproved." Modern science cannot explain everything, not by a long shot. Many things that were once thought impossible, or within the realm of magic, have since been scientifically proven. Remember that one time, the majority of the scientific community considered heavier-than-air flight to be theoretically impossible. Germ theory was once officially pooh-poohed as a superstitious fantasy. The idea of home computers was considered almost laughable science fiction until they started appearing in the marketplace. Hypnotism was once considered an occult, magical phenomenon, but today it is considered a valid and useful scientific and therapeutic tool. (The science of hypnosis is covered in chapter 2.)

Many New Agers point to these past mistakes of the scientific community as support of their beliefs. But just because something has not been disproved is a poor reason to believe that it must be true. This reasoning is merely superstitious thinking. But always remember, it is equally superstitious to take the position that just because something has not been proved true, it must be false. This is so even when the questionable idea is in opposition to current scientific theories. The history of science is filled with theories that were once widely held as unquestionably true and were later discredited.

The truly scientific position on most of these New Age ideas is, "No one knows for sure." They are neither proved nor disproved. Everyone has a right to their own opinion. It is one thing to say, "I believe *X* is true" or "I believe *Y* is false," but it's something altogether different to insist, ''*X* is absolutely true'' or ''*Y* is unquestionably false,'' and to state directly or indirectly that, ''Anyone who disagrees with me is wrong and a fool!'' People on either end of the debate over New Age concepts tend to fall into just such a superstitious, antiscientific viewpoint. Unfortunately, even many respected scientists with excellent credentials fall into this trap (usually on the negative end—''That idea is false!''). But regardless of their academic background and past experiences and accomplishments, anyone who adopts such an attitude is being superstitious. Such dogmatic attitudes (pro or con) are the very opposite of science.

Many New Age concepts are quite intriguing and certainly deserving of research. Without extensive experimentation, no one will ever know whether any of these ideas are really true or false. The electronic projects described throughout this book are to help you perform your own experiments into these fascinating areas on the edge of science. In no way does building one of these projects imply belief or disbelief in the underlying concepts of the project. At the very least, these projects are fun to work and experiment with.

By necessity, these projects are rather crude compared to what is available to modern laboratory researchers. Don't expect to conclusively prove or disprove anything with any of these projects. The purpose of these projects is to explore various unusual phenomena on a casual, hobbyist level. However, don't be surprised if you find that the results of one or more of your experiments cause you to reconsider some of your beliefs. True science works that way sometimes. Ultimately, science is not so much a set of absolute answers as the systematic raising of questions.

New Age beliefs and concepts

New Age concepts range from simple variations on long-standing philosophical ideas to beliefs in some fairly far-out phenomena. Some New Agers are reasonable, intelligent, well-educated people, while others are unrealistic fanatics hovering on the lunatic fringe. Of course, it is always an individual judgment call as to just what is part of the lunatic fringe and what is reasonable. Many, if not most, New Agers consider some New Age ideas to be rather ridiculous or, at least, insignificant.

In this section, I will introduce some of the more common New Age concepts. In each case, you are fully entitled to your own opinion. But please realize that it is your opinion and it isn't necessarily better than anyone else's. With the concepts discussed here, you can bet that all of the facts aren't in yet.

Probably the most common element throughout the entire New Age movement is a belief in the essential connectedness of all life, or even of all the universe. According to many New Age philosophies, everything is truly alive (although not necessarily in a strict biological sense). Crystals (discussed in chapter 9) are considered to be a life form, albeit of a very different sort.

Some religiously oriented New Agers believe that everything that exists is really part of God, or a universal spirit, or a cosmic mind. Less religious New Agers might hold similar ideas, though they might be inclined to avoid words such as "God." When a New Ager says, "I am God," he or she does not mean it in the sense that an Egyptian pharoah claimed to be a god. The meaning here is that because everything is part of God, each of us is God in the same way a drop of sea water is the ocean. In its fundamental makeup, the drop of sea water is the ocean in miniature. The ocean is a little less complete if that drop is removed. But, of course that drop is far from being the entire ocean, and the ocean can probably get along just fine without one individual drop. But if too many drops are removed, then the ocean is going to be noticeably depleted.

Because of this underlying assumption of universal interconnectedness, New Age ideas place great emphasis on brotherhood, service, and the Golden Rule. Ecology issues are often important to New Agers for similar reasons. Many New Agers believe, "What I do unto others is done unto me, because there is no real distinction between myself and others." "Negative" controlling cults such as Satanism are not usually considered a part of the New Age movement, although there is some overlap. Many New Agers have a strong interest in the occult, but most New Age teachings strongly criticize misuse of the occult and point out that such misuse inevitably has negative consequences for the abuser.

New Agers believe humanity is in the process of spiritual evolution. We are not as advanced or fully developed as we could be. This idea certainly doesn't seem very far-out. Is it reasonable to assume that the human race has met its full potential? Of course, there is considerable room for debate on just what the potential might ultimately entail.

Unlike the biological theory of evolution where simple entities are followed by more complex and sophisticated entities, the New Age spiritual evolution concept suggests that we are returning to a former lost state of full potential, or a rerecognition of what we truly are, though we have, as a species, forgotten the truth we once knew.

Because this spiritual evolution is assumed to be continuous, in a sense we are always entering a New Age or a new step in our development. However, many New Agers believe this is a critical time, and humanity as a whole is about to take or is in the process of taking a major step forward. Hence, we are on the cusp of a New Age in our spiritual development.

As a somewhat related issue, many New Agers believe in reincarnation in some form. Usually this does not include transmigration to other species, as in the Hindu religion. Because everything (according to New Age beliefs) is living and part of the universal spirit (or God), then death is logically impossible in any real sense (although biological bodies certainly stop and decay). Because human souls are not yet fully developed, reincarnation through multiple lives is thought to give the soul a chance to learn additional lessons and progress toward its state of full potential.

These New Age ideas are mostly within the realm of philosophy, and science can't legitimately say very much about them. How could anyone devise empirical experiments to test whether we are really part of God or a universal cosmic mind? It would be very difficult to prove or disprove such ideas scientifically.

Other New Age concepts deal more directly with how things work (at least within the physical realm). These ideas postulate cause-and-effect relationships, just like more traditional scientific hypotheses and theories, although philosophically they are far removed. It is possible (and desirable) to devise empirical experiments to test such New Age theories, and that is precisely what we do with the projects presented in later chapters.

Because of the nature of this book, we concentrate on the physical phenomena of the New Age. However, the underlying philosophies might be needed to make sense of some of the ideas in later chapters. Unfortunately, I can't go into enough detail to do these ideas justice. Many books on these concepts are available. I am not asking the reader to believe or accept any of these ideas. (Nor am I consciously revealing what I personally believe about these things.) I'm just outlining the base concepts that are needed to examine the New Age theories of the various phenomena we will explore with the projects in the later chapters of this book. You definitely do not need to believe the theories to experiment with and even be fascinated by the phenomena, whatever the true cause (or causes) may be.

Bear in mind that some people who consider themselves (or are considered by others) to be New Agers might strenuously disagree with what I've said here. I am just trying to briefly summarize some of the core beliefs that appear to be most common throughout the New Age movement. No single belief is universal because "New Age" is such a broad, ambiguous label.

Healings

Many New Agers (but not all) believe the physical body is not truly real or is of a lesser reality than the spirit or mind. This belief is the reverse of the traditional scientific/medical model, which tends to place greatest importance on what can be touched or directly and precisely measured. But the New Age shift of emphasis to the unseen and the immeasurable is not completely alien to science. Psychology, for one example, deals with similarly nebulous subject matter.

In most New Age theories of health, the body and its condition can be affected or even fully controlled by the mind or spirit. Disease is often considered an illusion; it can be counteracted mentally or spiritually. Nonmedical healing is a major part of the New Age. Treatments range from the use of herbs (actually a form of the traditional medical approach, just utilizing different substances) to unusual physical treatments such as acupuncture. Sometimes the physical element is eschewed almost entirely, and healing is attempted strictly through mental or spiritual means, perhaps with the laying on of hands. Some healers claim to be able to enact healings over considerable distances.

While none of these healing techniques has been scientifically proved effective, some of them do appear to work in some cases. (Even traditional medicine is not effective in all cases.) The scientific "proof" against most New Age healing techniques is actually quite weak and has more to do with preformed assumptions than true scientific research. Clearly, further research into these techniques is needed. They might work on a psychosomatic level, or they might be related to laws of nature we have not yet discovered. I will discuss New Age ideas of healing throughout this book, especially in chapters 2, 3, 5, and 7.

Crystals

Crystals are very popular among New Agers, but there are wide-ranging beliefs about crystals. Some New Agers don't think of crystals as much more than lovely, symbolic objects. Others believe that crystals contain a great deal of power (usually, but not always, psychic in nature) that can be tapped by a knowledgeable human. Still other New Agers take a sort of middle ground, holding that the crystal itself is not necessarily the source of the power, but a powerful focusing device for a person's psychic or mental energies. Chemically these crystals are no different than those used in electronic crystals, although New Age crystals tend to be much larger. Crystals are such an important and widespread concept in the New Age movement that I devote an entire chapter to them.

UFOs

Unidentified flying objects (UFOs) are one of the more controversial areas in the New Age movement. Some New Agers reject the idea of UFOs entirely and insist it is not a part of New Age at all. Others believe very strongly in UFOs. They believe UFOs are spaceships from some other planet or another dimension. Within the New Age movement, unlike many earlier beliefs about UFOs, these aliens are usually assumed to be benign and even actively beneficial. They are trying to help our spiritual growth. Many believers yearn for personal contact with these friendly aliens. Some claim they have already had (or continue to have) such contact.

Other New Agers suggest that UFOs are spiritual entities of some sort, perhaps related to angels. Because they are not physical objects, this theory would explain why they rarely show up on tracking equipment. Because it sidesteps normal scientific measurability, this theory would be very hard to prove or disprove in any practical laboratory.

Another popular New Age theory about UFOs is that they are a psychic phenomenon projected from the minds of groups of human beings or even humanity as a whole. Whether UFOs are strictly a hallucination or if they acquire some sort of external "reality" is an irrelevant question within this concept. What counts is what the projections mean. In other words, UFO sightings are primarily a symbolic rather than a literal phenomenon. This framework of ideas is often linked with theories from Jungian psychology (which is often studied by New Agers).

In one sense it is rather ridiculous to say, "I don't believe in UFOs." If it's up there and you don't know what it is, then it is a UFO. Even if it is actually a weather balloon, if you don't know that, it's still a UFO. Notice that the projected symbolic interpretation still applies, even if the UFO sighting is later proved to be a weather balloon. The psychic and emotional effects on the sighter remain the same. I briefly touch on UFOs again in chapter 8.

ESP

Extrasensory perception (ESP) is not strictly a New Age concept, although many New Age concepts do incorporate ESP. The term *ESP* refers to a number of phenomena that involve receiving or transmitting information without the use of the five ordinary senses—sight, hearing, smell, taste, and touch. For example, *telepathy* is direct, mind-to-mind contact. Person A can somehow detect and know what person B is thinking. *Telekinesis* is the ability to move objects without touching them using the force of the mind. *Clairvoyance*, strictly defined, means seeing at a distance. This phenomenon might involve feeling what is happening to a physically distant loved one, or it might involve foretelling the future.

Most aspects of ESP sound rather far-fetched to most of us. But consider the fact that scientists know that the average person actually uses less than 10% of his or her brain. What is the rest of the brain for? Could nature really be that wasteful? Is 90% of the brain devoted to automatic functions or emergency backup? Looked at this way, it seems rather unlikely and far-fetched to suggest that we don't have additional mental abilities we are not aware of. This reasoning does not prove that ESP exists, not by a long shot. But it does suggest that such things are theoretically possible or, at least, not conclusively disproved.

Experimenting with ESP (especially clairvoyance and telepathy) is covered in chapter 4.

Biological functions

Many of the body's operations are normally automatic, permitting us to devote our conscious minds to other things. For example, we normally don't have to think about breathing, though we are doing it almost constantly (unless we are deliberately holding our breath). The body automatically moves the necessary muscles and adjusts the amount of air we take into our lungs with each breath. Breathing adjustments happen without our thinking about them. The task is performed by an unconscious part of our brains. But you can use your conscious mind to override the automatic functioning. You can hold your breath or choose to breathe deep or shallow, fast or slow. Breathing is therefore a semiautomatic function.

Other body functions are normally considered to be purely automatic and outside the reach of our conscious minds. Examples of purely automatic body functions include heartbeat, digestion, and body temperature. Through biofeedback (see chapter 3), people apparently can learn to obtain a surprising amount of conscious control over such supposedly automatic body functions. New Agers use biofeedback to aid in meditation and to increase mental, and even psychic, powers.

A related concept is that of biorhythms. A great many things in nature operate according to regular repeating patterns or rhythms. Is our overall well-being and ability to function tied into a cyclic rhythm? According to the biorhythm idea, we have three regular cycles, each with its own rhythm or repetition rate. These cycles are physical, intellectual, and emotional. I will discuss biorhythms in chapter 6.

New Age and quantum physics

As mentioned earlier in this chapter, many New Age concepts have received surprising support by the developments and discoveries of quantum physics. We can't begin to fully cover quantum physics here. We will merely glance at just a few of the relevant concepts very briefly.

Don't worry if you feel you don't understand this subject. It is very complex and truly mind-boggling. In fact, many scientists working in this area have said (semijokingly) that anyone who thinks they understand quantum physics is demonstrating that they don't understand quantum physics at all. In a sense, the subject is inherently beyond human understanding. In many ways, it doesn't appear to make any sense at all. But the experiments also seem to work out, fulfilling the peculiar predictions of these strange quantum theories.

Quantum physics deals primarily with subatomic particles that combine to form protons, electrons, and neutrons, which, in turn, are the basic building blocks of all atoms. On the quantum level, the ordinary laws of physics break down. The behavior of a subatomic particle is often totally unpredictable. This unpredictability is on the theoretical level, and by all indications, it is not just a limitation of existing procedures. For example, you could determine how fast a particle is moving, but then you'd have no idea of where it is at any given instant, or you could determine exactly where it is at a specific instant, but then you would be completely unable to know anything at all about its movement. It is theoretically impossible to simultaneously know both the position and the velocity of a quantum particle. At best, you can compromise and measure the approximate speed and the approximate location. The more we know about the particle's position, the less we can know about its velocity, and vice versa. This uncertainty is not a limitation of measurement techniques, but a fundamental limitation of the nature of the particle itself. Strange as it sounds, if we know a particle's exact velocity, then that particle is simultaneously everywhere at once.

No, you can't cheat and measure the velocity of a particle then measure its position at a later time or have someone else make the second measurement. Somehow, the very act of making a measurement actually affects the quantum particle's very nature. Once the velocity of a quantum particle is precisely measured, it really has no definite position at all. Strange as it sounds, the particle is literally everywhere at once.

If two quantum particles are in contact with each other and are then separated, they might, under certain conditions, act as though they are still in full contact with each other. If something acts on particle A, particle B instantaneously shows the same effects, even if the information of what happened to particle A had to travel faster than the speed of light (which is physically impossible) to reach particle B so it can react accordingly. Such simultaneous reactions have been observed many times in laboratories, even though they are impossible to explain and apparently violate the most fundamental laws of classic physics. The results of quantum experiments should be impossible, yet they definitely occur. Could such quantum effects possibly begin to explain such phenomena as clairvoyance and telepathy?

In quantum physics, it is clear that all subatomic particles are interconnected in ways not yet fully understood. Because everything is made up of subatomic particles, this means that everything that exists is interconnected on some level. This theory is a rewording of the basic New Age concept about the unity of the universe, which was discussed earlier.

New Agers and esoterics throughout the ages have long asserted that all matter is really just spirit, or energy, vibrating at specific frequencies or patterns. Physicists studying subatomic particles have almost conclusively confirmed this idea, although they use different terminology. New Agers say all matter is really energy. Quantum physicists say that all matter is made up of tiny, mysterious particles known as *quanta*, which are not really matter at all, but packets of contained energy. Are the differences here really all that extreme?

At the very least, quantum physics is so bizarre and impossible that many of the claims of New Age phenomena seem almost reasonable by comparison. Keep an open mind and see what you can discover experimentally. Always remember what Hamlet said, "There is more in heaven and earth than is dreamed of in your philosophy"—or in our science.

If nothing else, the experimental equipment described in the later chapters of this book is fascinating, and you should have a lot of fun with it. In fact, several of these projects should be just great at parties.

❖ 2 Meditation and hypnosis aids

PROBABLY THE MOST COMMON OF ALL NEW AGE PRACTICES IS meditation. There are a number of different approaches to meditation, but they all attempt to still the mind. Normally the human mind is extremely active, jumping from thought to thought in a haphazard fashion. Much of this mental energy is wasted because it doesn't accomplish anything. Most of our thoughts are rather worthless and meaningless. They are wasted mental energy. In meditation the goal is to stop (or at least attenuate) this constant stream of stray thoughts. The result of this mental quieting is relaxation on the physical, emotional, and spiritual levels. Your mental energy is focused and brought more fully under conscious control, rather than being randomly scattered.

Benefits of meditation

The relaxation offered by meditation can reasonably be expected to aid in mental health. Meditation can also aid in physical health and healing. This is not a magical claim, and there is quite a bit of scientific evidence to support it. After all, meditation reduces stress, and most medical practitioners now recognize that stress lowers our resistance to disease and retards healing. So it should not be surprising that stress-reducing meditation results in improved physical health. There are also spiritual and religious aspects to meditation, but I won't go into them here.

When most people attempt meditation they find that they cannot empty their minds of all thoughts. It is very, very difficult to truly think of nothing at all. Some dedicated meditators can accomplish it for extended periods, but only after years of practice. Most of us, however, need to resort to some sort of "trick" to achieve the effects of meditation. Many schools of meditation emphasize the use of a mantra (or something similar). A *mantra* is a special word or sound, repeated over and over during meditation. The repetitious nature of the mantra occupies the mind sufficiently to reduce distracting thoughts. When a stray thought is noticed, the meditator brings himself back into focus by thinking about his mantra.

A similar technique would be to stare at some object or spot, letting the visual image occupy the mind. This is where the stereotypical (and often comical) image of the yogi meditating on his navel comes from. There is nothing special about the navel. It's just a convenient spot to focus on to aid in meditation. The yogi is not really meditating on his navel in the sense of thinking about it or contemplating it. He is simply focusing on this handy, self-contained spot to clear his mind during meditation.

With modern technology, we can develop more sophisticated meditation aids than our bellybuttons. Sight or sound can be used. The advantage of electronic meditation aids is that they have specific frequencies. By synchronizing the brain's waves with a compatible and consistent external frequency, a tremendous degree of mental focusing can be achieved in a relatively short time.

Hypnosis

Closely related to meditation is hypnosis. Hypnosis is not, in itself, a New Age concept, but it is often used as a tool by New Agers, and it has the same undeserved "occult" reputation of many New Age concepts. Because it is so closely related to meditation and can use the same electronic aids, I will briefly discuss the subject here.

Few subjects have been so misunderstood and misinterpreted by both scientists and the general public as hypnosis. For a long time the accepted viewpoint was that there really was no such thing as hypnosis. It was just a hoax or a mistake. Few scientists today continue to hold this derogatory view of hypnosis. The existence of hypnosis as a real phenomenon has been well-established, though there is still some debate over exactly what it is and how it works. In the following discussion, I will concentrate on the most commonly held theories, which seem the most reasonable in light of the existing evidence.

Early practitioners of hypnosis didn't understand what was going on when someone was hypnotized. Their mistaken (and often naive) theories did much to establish the questionable status of hypnosis.

Hypnosis was discovered (not invented) by a man named F.A. Mesmer, and his techniques were often called "mesmerism." His theories were so off-base that this name soon developed a very negative connotation, and it is rarely used today. If you do encounter the word somewhere, it's just an obsolete synonym for the word "hypnosis."

Mesmer referred to his discovery as "animal magnetism" and visualized it as a power originating within the hypnotist and projected out to the subject being hypnotized in the form of some sort of mysterious "magnetic" rays. The hypnotized subject was then fully under the control of the hypnotist.

This imaginative theory has long since been disproved, though elements of it continue to hang on in the minds of the general public. Hypnosis is not the result of any "power rays" emitting from the hypnotist or anywhere else. It is simply an altered state of consciousness, differing from ordinary, waking consciousness. Sleep, daydreaming, and even meditation are other altered states of consciousness. We aren't talking about anything mysterious or magical here.

In fact, hypnosis is very natural and much more common than most people realize. Have you ever watched a movie and cried at a sad part of the fictional story being presented? Have you ever been so engrossed in reading a book that you didn't hear what someone said to you? Have you ever been caught up in the words of a dynamic public speaker, then afterwards thought it over and decided you didn't agree with the speaker at all? Each of these experiences (and many others like them) is a form of natural hypnosis. There probably isn't a single person on earth (except possibly the severely brain damaged) who has never been hypnotized.

Hypnosis is often considered similar to sleep. In fact, many (and perhaps most) hypnotists use phrases such as, "You are getting sleepy." The word "hypnosis" comes from the name of the mythological Greek god of sleep—Hypnos. Superficially there are a lot of obvious similarities between hypnosis and sleep. In both cases, the subject is usually physically relaxed, breathing slowly and deeply. Both the hypnotized person and the sleeping person seem relatively unaware of stimuli within their environment.

But when we look closer, we find hypnosis and sleep are very different. Brain wave patterns of a hypnotized person do not resemble those of sleep. In fact, they suggest a heightened level of consciousness, not unconsciousness. A sleeping person is more or less oblivious to all sensory stimuli (sights, sounds, etc.). The conscious mind is essentially shut off. In hypnosis, however, the subject has heightened awareness and greater focus and selectiveness to external sensory stimuli. Some stimuli (usually the hypnotist's voice) receive great concentration, while others (usually normal environmental cues) are ignored.

The conscious mind is not shut off during hypnosis, but it relaxes its normal level of control and lets more of the unconscious mind emerge. The unconscious mind tends to be highly suggestible, which is why hypnotic suggestions or "commands" work. However, contrary to popular belief, a hypnotized person is not fully suggestible and will not obey all hypnotic commands like a mindless zombie. If there is something you really don't want to do (perhaps because of deep set moral convictions or potential danger), you won't do it under hypnosis. Most people cannot be hypnotized to commit murder or take their clothes off in public. If such a command is given by the hypnotist, one of several things will happen: (1) the suggestion will be ignored; (2) the hypnotized subject will say, "No" (usually in a flat, unemotional tone); or (3) the subject will come out of the hypnotic state, either returning to ordinary consciousness or simply falling asleep.

No, it will not work to give a hypnotic suggestion so the subject thinks he or she is in a different environment. Part of the conscious mind is still active and aware of the true environment at all times during hypnosis, even though this information is ignored for the most part. For example, a stage hypnotist might suggest that a woman is alone in her bedroom. She will go along with the suggestion, acting just as if she really was alone in her bedroom. But if he suggests that she take her clothes off for a shower, most women will not comply with that hypnotic "command."

A hypnotist does not control the hypnotized subject. The appearance of control is due to the voluntary suggestibility of the hypnotized subject. No hypnotist can make anyone do anything he really doesn't want to do. However, he might do something he didn't think he wanted to do, but had subconscious urges to do. Someone on a strict diet might well eat a piece of cake under hypnosis when he'd faithfully say no to that cake without hypnosis. Part of him really wanted to eat the delicious cake, even though he consciously "knew better."

The circumstances surrounding the hypnosis can also be of significance in determining which suggestions will be accepted and which will be rejected. For instance, let's consider two common types of hypnosis—a stage hypnotist's show and a psychotherapist using hypnosis to treat a patient. If a stage hypnotist tells a subject to cluck like a chicken, he will probably do so. After all, he agreed to "perform" as part of the show when he allowed himself to be hypnotized on stage. But if a psychotherapist told a hypnotized patient to cluck like a chicken, it is probable that the patient will quite reasonably ask, "Why?" The hypnotized patient's conscious mind, which is willingly inactive during hypnosis, would realize that this suggestion doesn't sound like a reasonable part of the therapy and is therefore inappropriate. Similarly, if the stage hypnotist asked a subject to relate the intimate details of some deeply personal trauma, he wouldn't cooperate, even though the psychotherapist would probably get a detailed report from the same suggestion. Notice that at all times, the hypnotist must play by the rules of appropriateness laid down by the hypnotic subject, not the other way around.

A hypnotist is not so much a controller as a guide. The hypnotized person does most of the work himself. There is good cause to believe that ultimately all hypnosis is really self-hypnosis. The hypnotized person will listen to and accept the external hypnotist's suggestions only if he has given himself the self-hypnotic suggestion to do so. The hypnotist never does and never can take complete control of any hypnotic subject.

Simple audio hypnotizer

One of the easiest ways to induce a hypnotic trance (or a meditative state) is with a steady, regularly repeating sound. That is the purpose of the two projects presented in this section.

The circuit for our first audio hypnotizer project is shown in Fig. 2-1. If you have worked much with electronic circuits before now, you will undoubtedly recognize this circuit. This project is

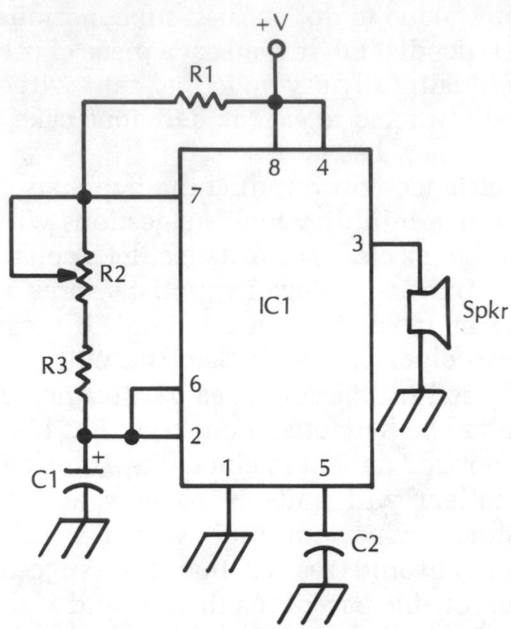

Fig. 2-1 *Simple audio hypnotizer project.*

simply a low-frequency astable multivibrator built around a common 555 (or 7555) timer IC.

The output pulse rate is very low—about once every two seconds, depending on the setting of the potentiometer (R2). Each time the timer's output changes its state (from low to high, or vice versa), a clicking sound will be heard from the speaker.

The suggested parts list for this project is given in Table 2-1. You might want to experiment with alternate values. The basic pulse rate of the astable multivibrator is determined by the formula

$$F = \frac{1.44}{((R_1 + 2(R_2 + R_3))C_1)}$$

The speaker will emit two clicks per output pulse—one for the low-to-high transition, and one for the high-to-low transition. It is not possible to achieve a true square wave with this particular circuit, but by making the value of R1 large in comparison with the combined values of R2 and R3, the output waveform will be fairly close to the 1:2 duty cycle of a true square wave and the clicks will be fairly evenly spaced. On the other hand, you might find it more effective to use a relatively small value for R1

Table 2-1 Parts list for the simple audio hypnotizer project of Fig. 2-1.

IC1	7555 timer (or 555)
C1	10-μF, 35-V electrolytic capacitor
C2	0.01-μF capacitor
R1	100-kΩ, 0.25-W, 5% resistor
R2	50-kΩ potentiometer
R3	33-kΩ, 0.25-W, 5% resistor
SPKR	Small loudspeaker

and a larger one for R3 to create an asymmetrical click pattern, like "click-click—click-click—click-click . . ." with a definite pause separating each pair of clicks.

The parts list recommends the following component values:

- C1 10 μF (0.000 01 F)
- R1 100 kΩ (100 000 Ω)
- R2 (potentiometer) 50 kΩ (50 000 Ω) maximum
- R3 33 kΩ (33 000 Ω)

Being a potentiometer, R2 can be adjusted from its maximum value (50 kΩ) down to a nominal minimum of 0 Ω. (Actually a practical potentiometer will probably still have a measurable resistance at its minimum setting, but we can ignore that here.) Using these components, with R2 set for its minimum value (0 Ω), we get the maximum output frequency, which works out to

$$
\begin{aligned}
F &= \frac{1.44}{((R_1 + 2(R_2 + R_3))C_1)} \\
&= \frac{1.44}{((100\,000 + 2(0 + 33\,000))0.000\,01)} \\
&= \frac{1.44}{((100\,000 + 2(33\,000))0.000\,01)} \\
&= \frac{1.44}{((100\,000 + 66\,000)0.000\,01)} \\
&= \frac{1.44}{(166\,000 \times 0.000\,01)} \\
&= \frac{1.44}{1.66} \\
&= 0.87 \text{ Hz}
\end{aligned}
$$

$$T = \frac{1}{F}$$

$$= \frac{1}{0.87}$$

$$= 1.15 \text{ seconds}$$

There will be one complete pulse (two clicks) about once a second. This is the fastest available click rate from this set of component values.

If we increase the setting of R2 to its maximum value (50 kΩ), we get the circuit's minimum output frequency:

$$F = \frac{1.44}{((R_1 + 2(R_2 + R_3))C_1)}$$

$$= \frac{1.44}{((100\,000 + 2(50\,000 + 33\,000))0.000\,01)}$$

$$= \frac{1.44}{((100\,000 + 2(83\,000))0.000\,01)}$$

$$= \frac{1.44}{((100\,000 + 166\,000))0.000\,01)}$$

$$= \frac{1.44}{(266\,000 \times 0.000\,01)}$$

$$= \frac{1.44}{2.66}$$

$$= 0.54 \text{ Hz}$$

$$T = \frac{1}{F}$$

$$= \frac{1}{0.54}$$

$$= 1.85 \text{ seconds}$$

It now takes almost 2 seconds for every output pulse (pair of clicks). Notice that the clicks will be more closely spaced this time, giving a click-click-pause effect. This happens because we have increased the resistance of R2 without proportionately increasing the value of R1.

In experimenting with this circuit, you might want to try larger or smaller values for resistors R1 and R3. You might also try substituting different values for capacitor C1. Increasing any of these values will decrease the frequency, slowing down the clicks. Decreasing any of the component values, of course, will have the opposite effect of increasing the speed of the clicks. If the frequency is increased above about 5 Hz, you will no longer be able to hear the clicks very well, if at all. Increasing the frequency above 30 Hz will result in a constant, steady tone from the speaker.

Improved audio hypnotizer

A slightly improved circuit for the audio hypnotizer is shown in Fig. 2-2. The suggested parts list for this project appears in Table 2-2. Notice that IC1 and its associated components are set up as pretty much the same circuit as the one we were just working with (Fig. 2-1). IC2 and its associated components form a second astable multivibrator set up for an audio-range tone. Using the component values suggested in the parts list, this tone will have a frequency of about 1 kHz (1 000 Hz).

The outputs from these two astable multivibrators are fed through an AND gate (IC3). Actually the AND gate is made up of a NAND gate followed by an inverter (a second NAND gate with its input leads shorted together). Electrically the effect is the same as a single dedicated AND gate. This seemingly more-complex arrangement of gates is used because NAND gate ICs are usually more readily available and inexpensive. You could substitute a dedicated AND gate if you prefer.

The output of an AND gate can be high if and only if all of its inputs are high. This means that in this circuit, the tone can pass through to the speaker only when the output of IC1 is high, which it is for only a brief period in each low-frequency cycle. You will hear a brief burst of tone once every couple of seconds. The tone bursts will be very regular and evenly spaced. Adjust potentiometer R2 to control the rate of the tone bursts. Potentiometer R5 is used to adjust the frequency of the audio tone. Set these controls to suit your needs and achieve the best results.

An audio hypnotic/meditative aid like these projects will work best if other audio and visual stimuli are minimized as much as possible. Try meditating or performing hypnosis in a quiet, fairly dark room. Better results might be obtained if the subject closes his eyes or wears a blindfold of some sort (a simple thick cloth laid over the eyes will do).

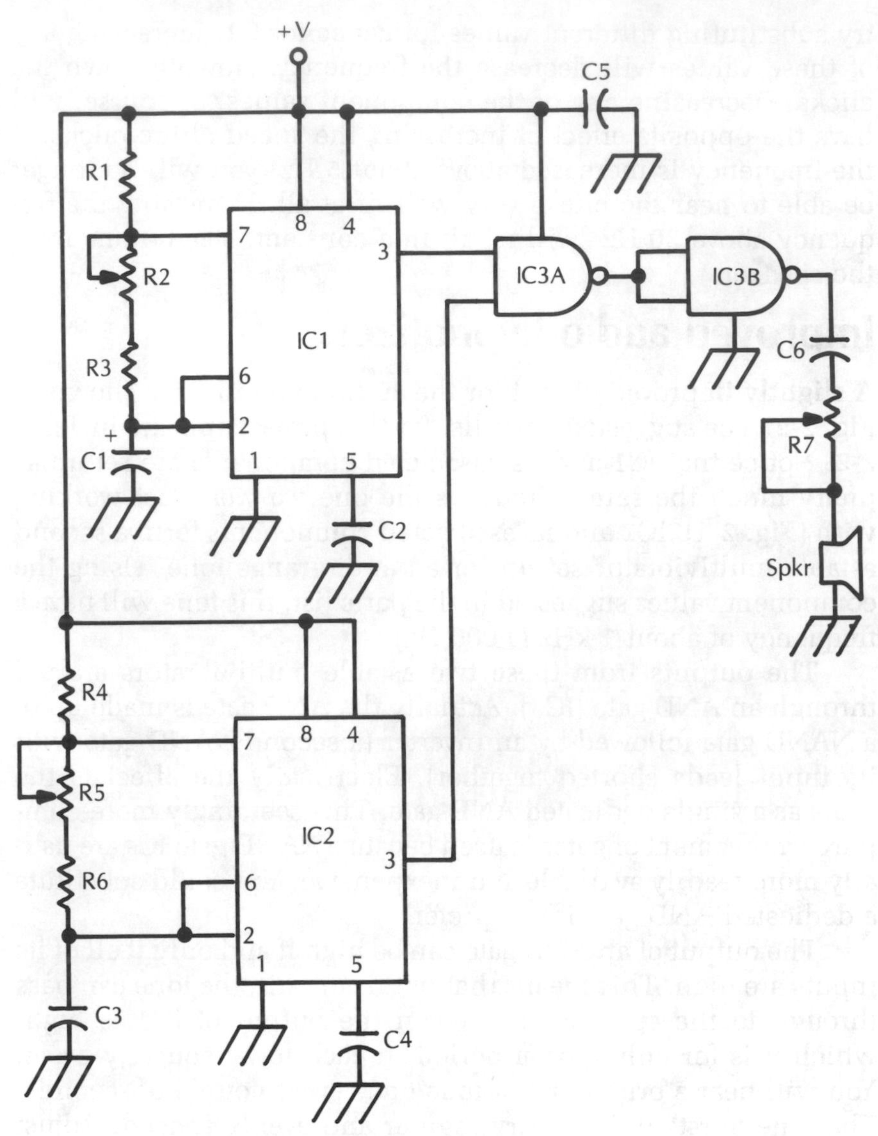

Fig. 2-2 Improved audio hypnotizer project.

Visual hypnotic aid

Some meditators and hypnotists prefer a visual rather than an audio focus. It is no problem to adapt the basic ideas from the preceding projects to a visual output. It's simply a matter of substituting an LED for the speaker at the output.

Figure 2-3 shows the schematic diagram for a simple visual hypnotic aid circuit. Compare this circuit with the audio version

Table 2-2 Parts list for the improved audio hypnotizer project of Fig. 2-2.

IC1, IC2	7555 timer (or 555)
IC3	CD4011 quad NAND gate
C1	10-μF, 35-V electrolytic capacitor
C2, C4	0.01-μF capacitor
C3	0.022-μF capacitor
C5, C6	0.1-μF capacitor
R1	33-kΩ, 0.25-W, 5% resistor
R2	100-kΩ potentiometer (rate)
R3	47-kΩ, 0.25-W, 5% resistor
R4	22-kΩ, 0.25-W, 5% resistor
R5	50-kΩ potentiometer (pitch)
R6	10-kΩ, 0.25-W, 5% resistor
R7	500-Ω potentiometer (volume)

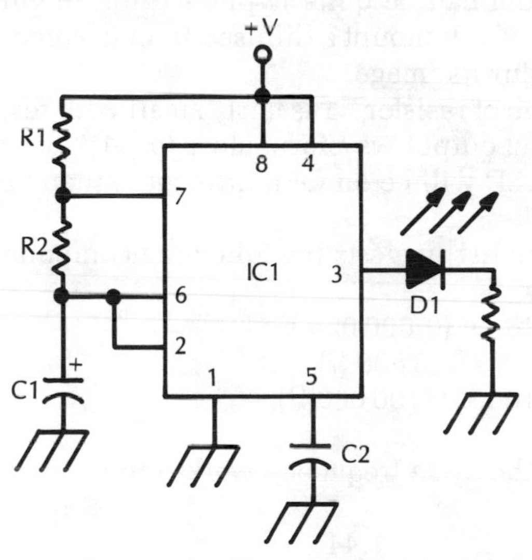

Fig. 2-3 Visual hypnotic aid project.

shown in Fig. 2-1. A suitable parts list for this project appears in Table 2-3. Feel free to experiment with alternate component values.

Once again, we have just a simple astable multivibrator circuit built around the common and popular 555 timer IC. If you choose, you might substitute a 7555 CMOS timer IC. The 555 and the 7555 are directly interchangeable, with identical pins. No other changes in the circuitry are required to accommodate this substitution. The chief advantage of using the 7555 in an application like this is that it consumes less current.

Table 2-3 Parts list for the visual hypnotic aid project of Fig. 2-3.

IC1	7555 timer (or 555)
D1	LED
C1	25-μF, 35-V electrolytic capacitor
C2	0.01-μF capacitor
R1	6.8-kΩ, 0.25-W, 5% resistor
R2	100-kΩ, 0.25-W, 5% resistor
R3	330-kΩ, 0.25-W, 5% resistor

When the output of the timer is in the high portion of each cycle, the LED will light up. During the low portion of each cycle, the LED will be dark. In operation, the LED flashes on and off at a slow, regular rate. Staring at the LED, particularly in a darkened room can be quite hypnotic. You might achieve even better results if you mount a diffused filter of some sort in front of the LED to blur its image.

The value of resistor R1 is fairly small with respect to resistor R2, giving the output waveform close to a 1:2 duty cycle (square wave). The LED will be on for a little more than one-half of each cycle.

The parts list suggests the following component values:

- C1 25 μF (0.000 025 F)
- R1 6.8 kΩ (6 800 Ω)
- R2 100 kΩ (100 000 Ω)

This means the cycle frequency works out to

$$
\begin{aligned}
F &= \frac{1.44}{((R_1 + 2R_2)C_1)} \\
&= \frac{1.44}{((6\,800 + 2 \times 100\,000)0.000\,025)} \\
&= \frac{1.44}{((6\,800 + 200\,000)0.000\,025)} \\
&= \frac{1.44}{(206\,800 \times 0.000\,025)} \\
&= \frac{1.44}{5.17} \\
&= 0.28 \text{ Hz}
\end{aligned}
$$

Taking the reciprocal, we find that the length of each cycle is equal to

$$T = \frac{1}{F}$$

$$= \frac{1}{0.28}$$

$$= 3.6 \text{ seconds}$$

The LED will be on for about 1.8 seconds, then off for the next 1.8 seconds, then the cycle will repeat. Experiment with different component values to try different flash rate times.

There is no real point in experimenting with alternate component values for capacitor C2. This component is included to improve the stability of the timer IC. It is not always essential, but it is cheap insurance against possible erratic circuit operation.

Keep the cycle frequency below about 3 Hz and preferably less than 1 Hz. At higher frequencies, the individual flashes will blend together due to the persistence of vision, and the LED will appear to be continuously lit, although at lower than normal intensity. The LED is still blinking on and off during each cycle, but at a rate too fast for the eye to detect the changes.

Dual LED visual hypnotic aid

You can achieve an even more hypnotic effect by using two LEDs instead of just one. The modified circuit is shown in Fig. 2-4, with the suggested parts list appearing in Table 2-4. Notice that the two LEDs are wired with opposing polarity. When one LED is on, the other will be off. As long as power is applied to the circuit, one and only one of the two LEDs will be lit. They will continuously switch back and forth. If the cycle frequency is too high, both LEDs will appear to be continuously lit because the eye can't catch the individual blinks. Mount the LEDs a few inches apart, so the subject needs to look back and forth from point to point. This can aid in the hypnotic effect.

Alternatively, you can use two LEDs of contrasting colors (red/green are available), mounting them as close together as possible to minimize eye movement. Try placing a diffused lens of some sort over the LEDs to blur their physical differences in position. The goal here is to make the two LEDs appear to be one, single light source that changes color at a regular rate.

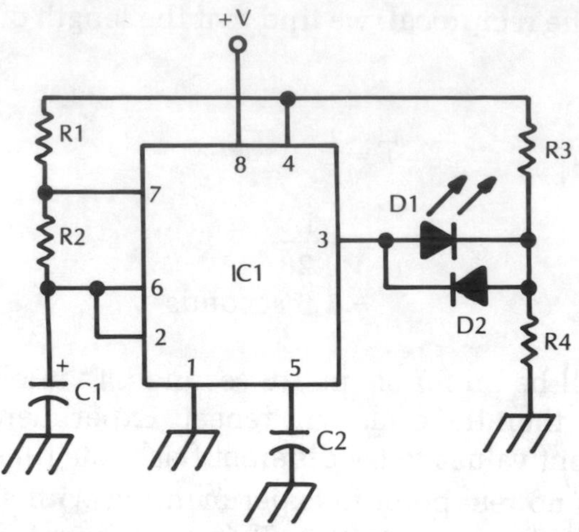

Fig. 2-4 Dual LED visual hypnotic aid project.

Table 2-4 Parts list for the dual LED visual hypnotic aid project of Fig. 2-4.

IC1	7555 timer (or 555)
D1, D2	LED
C1	25-μF, 35-V electrolytic capacitor
C2	0.01-μF capacitor
R1	6.8-kΩ, 0.25-W, 5% resistor
R2	100-kΩ, 0.25-W, 5% resistor
R3, R4	390-Ω, 0.25-W, 5% resistor

This would be an ideal application for a tristate LED. A tristate LED is nothing more than a pair of matched, reverse parallel LEDs in a single housing, as illustrated in Fig. 2-5. Usually one of the LEDs is red and the other is green. When a voltage of one polarity (the high portion of the rectangular-wave output from the timer) is applied across the tristate LED, it glows red. Revers-

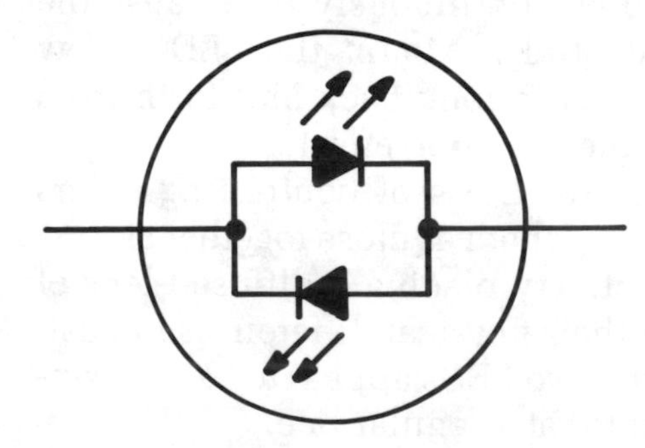

Fig. 2-5 A tristate LED is a nice touch in the dual LED visual hypnotic aid project of Fig. 2-4.

ing the polarity of the voltage (low portion of the rectangular-wave output from the timer) applied across the tristate LED causes it to glow green. If the applied frequency is too high for the human eye to distinguish the individual red and green blinks, the red and green glows will blend to form yellow.

Deluxe audio/visual hypnotic aid

The circuit shown in Fig. 2-6 combines the basic ideas of the projects we have been working with so far and offers both visual and audio stimuli. These stimuli are always in perfect synchronization, reinforcing their effects. The bursts of tone coincide with the red LED flashes.

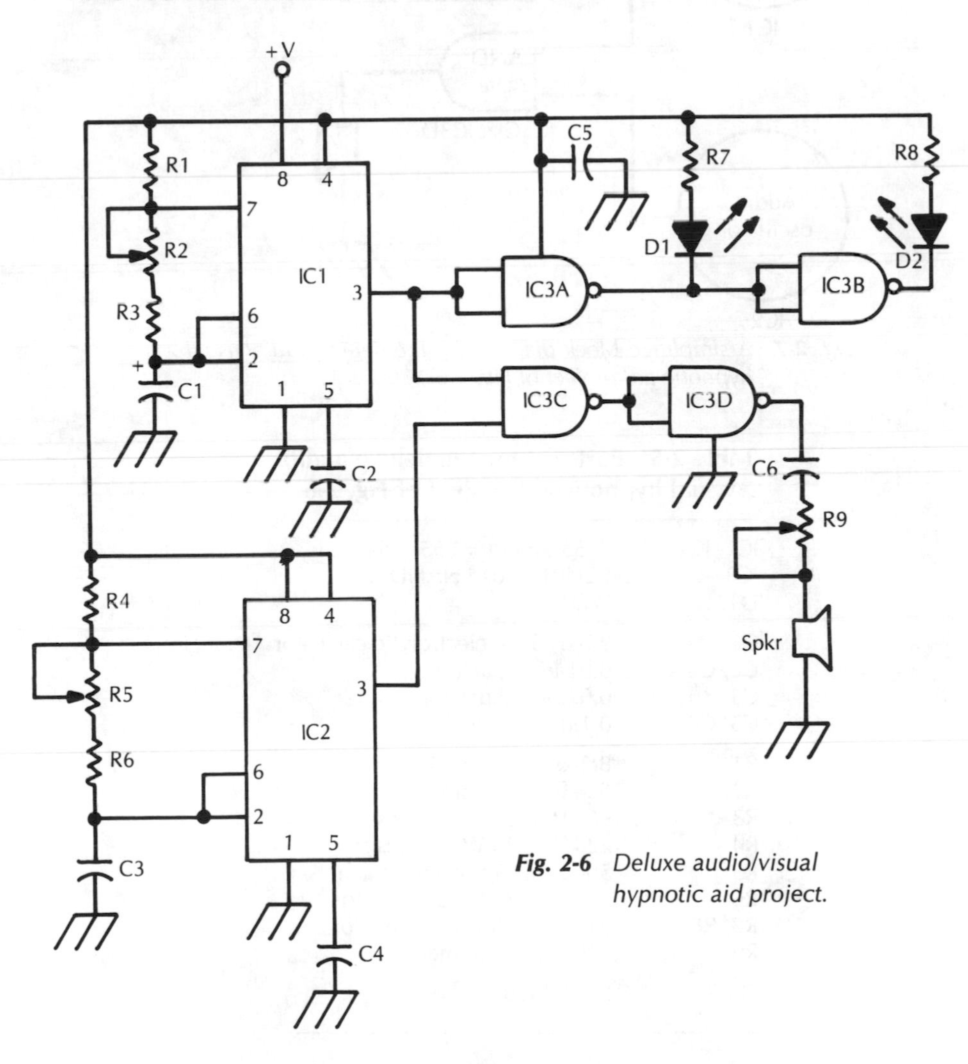

Fig. 2-6 Deluxe audio/visual hypnotic aid project.

This circuit is a little more complex than the earlier circuits presented in this chapter, but it is no trouble to understand if it is broken up into the functional elements indicated in the block diagram of Fig. 2-7. Basically we have the circuits of the earlier projects in this chapter blended together. A suitable parts list for this project is given in Table 2-5.

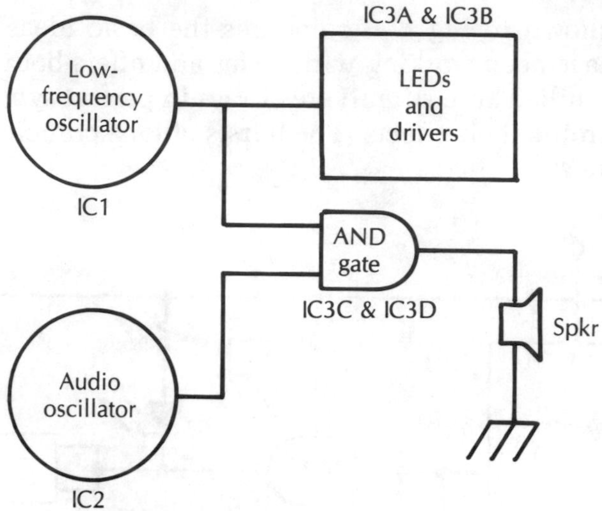

Fig. 2-7 *A simplified block diagram for the deluxe audio/visual hypnotic aid project of Fig. 2-6.*

Table 2-5 Parts list for the deluxe audio/visual hypnotic aid project of Fig. 2-6.

IC1, IC2	7555 timer (or 555)
IC3	CD4011 quad NAND gate
D1, D2	LED
C1	25-μF, 35-V electrolytic capacitor
C2, C4	0.01-μF capacitor
C3	0.022-μF capacitor
C5, C6	0.1μF capacitor
R1	8.2-kΩ, 0.25-W, 5% resistor
R2	50-kΩ potentiometer (rate)
R3	68-kΩ, 0.25-W, 5% resistor
R4	22-kΩ, 0.25-W, 5% resistor
R5	50-kΩ potentiometer (pitch)
R6	10-kΩ, 0.25-W, 5% resistor
R7, R8	330-Ω, 0.25-W, 5% resistor
R9	500-Ω potentiometer
SPKR	Small loudspeaker

Alpha glasses

There are a number of different types of brain waves, each with its own frequency. These frequencies vary somewhat from person to person. In ordinary waking consciousness, the most significant brain wave patterns are of a type known as beta waves. When you are very relaxed, your dominant brain waves are alpha waves. Alpha-wave frequencies run from about 7 Hz to 14 Hz. These rhythm patterns show up automatically within the brain when you are relaxed. Interestingly, it can work in the opposite direction. If your alpha frequency is flashed in your eyes, your brain will tend to synchronize or "resonate" with that visual stimulus, and you will enter a very relaxed alpha state. This project's purpose is to provide the alpha frequency visual stimulus. Remember, the exact alpha frequency varies from person to person, so you will need to fine-tune the exact flash frequency for each individual user in order to achieve the desired results.

The basic circuit for this project is shown in Fig. 2-8. As you can see, this is not a particularly complex circuit. A suitable parts list for this project is given in Table 2-6. Once again, the heart of the circuit is the ever-popular 555 timer IC, which is operated in the astable mode. In this particular application, I recommend using a 7555 chip, which is the CMOS version of the 555; but an ordinary 555 will work. The 7555 and the 555 are pin-for-pin compatible and don't require any different external circuitry when substituting one for the other.

The transistors serve as a driver amplifier to flash two LEDs (one for each eye) at the alpha frequency. This transistor driver is used to maximize the brightness of the LEDs, while minimizing the project's current drain. You can increase the LED brightness by reducing the value of resistor R11, but be sure to test the project with the recommended value first before making such a change. Too low a value for this resistor will result in painfully bright flashes. Some users might find the LEDs too bright even with the 15-Ω resistor suggested in the parts list. You can decrease the brightness of the LEDs to a more comfortable level by increasing the resistance of R11. I don't recommend making this resistor larger than about 33 Ω.

The exact transistor types used here aren't too terribly critical. Almost any low-power signal transistors will probably work fine. Transistor Q1 must be a pnp bipolar transistor and Q2 and Q3 should be npn bipolar transistors.

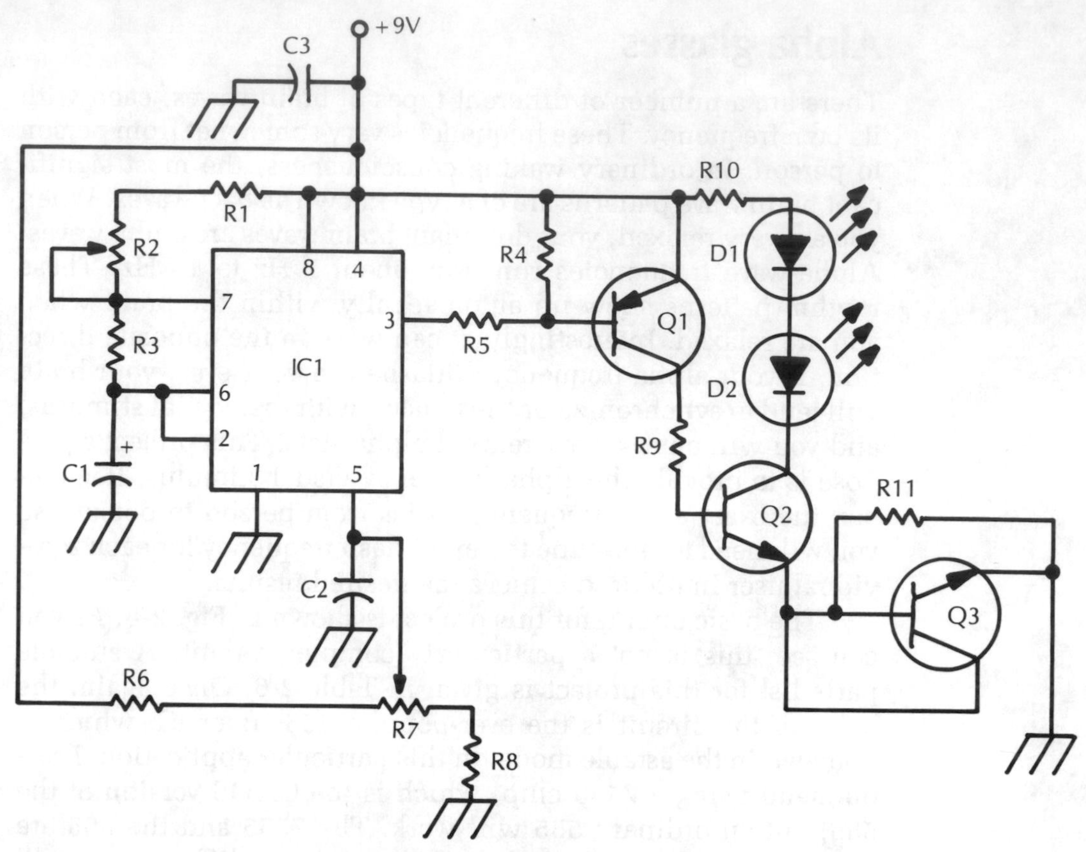

Fig. 2-8 *Alpha glasses project.*

Table 2-6 Parts list for the alpha glasses project of Fig. 2-8.

IC1	7555 timer (or 555 timer) (see text)
Q1	pnp transistor (GE21, HEP715, SK3025, ECG159, or similar)
Q2, Q3	npn transistor (GE20, HEP736, SK3122, ECG128, or similar)
D1, D2	LED
C1	10-μF, 25-V tantulum capacitor (see text)
C2	0.05-μF capacitor
C3	0.1-μF capacitor
R1, R9	3.9-kΩ, 0.25-W, 5% resistor
R2, R7	10-kΩ potentiometer (see text)
R3	1.8-kΩ, 0.25-W, 5% resistor
R4, R5	10-kΩ, 0.25-W, 5% resistor
R6	12-kΩ, 0.25-W, 5% resistor
R8	27-kΩ, 0.25-W, 5% resistor
R10	100-Ω, 0.25-W, 5% resistor
R11	15-Ω, 0.25-W, 5% resistor

Two potentiometers are used to control the frequency of the LED flashes. These can be either trimpots or front-panel controls, depending on how you intend to use the project. Potentiometer R2 is the coarse frequency control used to set the frequency over a range extending from a little under 7 Hz to a bit above 14 Hz, so the entire normal alpha range is covered. Potentiometer R7 can be used to fine-tune the actual frequency more precisely. The fine-tune control can alter the coarse frequency by up to 20%.

If just a single user is likely to be using the finished project, it might make more sense to use a trimpot for the coarse frequency control (R2) so the approximate frequency can be more or less permanently set. A front-panel (standard potentiometer) control can be used to fine-tune the actual frequency (R7).

The best frequency setting will be somewhat subjective. You will need to find it by trial and error. Slowly move each control (first the coarse frequency control (R2), then the fine-tune control (R7)) through its range until you find a setting that feels right. Often when you hit the best frequency for you, the effect will be quite dramatic and unmistakable. The flashing LEDs might suddenly seem to get brighter, or you might find your eyelids or the muscles around the eyes twitching slightly in step with the rhythmic flashes. It is not essential to set the frequency with exact precision, but adjust the controls for the strongest possible effect. Find a frequency that you seem to naturally synchronize with.

Because the signal frequency is moderately critical in this application and because we are concerned with a fairly limited range of usable frequencies, it is a good idea to use a timing capacitor (C1) with relatively high precision. A tantulum capacitor will do a much better job than a standard electrolytic capacitor. However, if you are only interested in crude experimentation, you might be able to get away with an electrolytic capacitor for C1.

Mount the LEDs outside a pair of swimming goggles—one LED in the center of each eyepiece. Enclose each LED within a plastic cap, such as those for pill bottles or film containers. If the end caps are clear, paint them black or some other dark color to shield against any external light when wearing the goggles. Make sure the cap completely covers each eyepiece lens. Block off any potential light leaks. This way you can get the full benefit of the project without needing to sit in a darkened room, as was the case for the visual hypnosis aids presented earlier in this chapter.

In use, keep your eyes open as much as possible. However, the closely placed light sources (the LEDs) should be bright enough to shine through your eyelids, so it will still work even with your eyes closed. You'll notice the best results are obtained when you are in a quiet area or with quiet, soothing music. So called "New Age" music is very suitable as a background for meditation. An audio hypnosis aid might be helpful, but it might be distracting if it is not well-synchronized with the light flashes within the goggles. These meditation goggles will also work well as an aid to induce a hypnotic state. Both meditation and hypnosis are relaxed mental states with strong alpha-wave activity.

This project can be adapted for use with theta waves instead of alpha waves. Theta waves are somewhat lower in frequency than alpha waves. While alpha waves are typically between 7 Hz and 14 Hz, theta waves have frequencies of about 4 Hz to 7 Hz. Theta waves are associated with a deeper stage of sleep or trance than alpha waves. Using theta waves is a good sleep aid for an insomniac. Theta waves are probably not a very good choice for meditation, simply because they are likely to cause you to drift off into sleep rather than into a true meditative state.

The only modification you have to make to the circuit so it will work with theta waves is to double the value of timing capacitor C1. This will cut the circuit's output frequency in half so it operates in the theta range.

WARNING

For most people, using a project like this is perfectly safe, but if you are an epileptic, do not attempt to use this project. It could induce a seizure. Even if you have not been diagnosed as an epileptic, be very cautious the first time you try your alpha goggles. Make sure someone else is present in case of an emergency. If you notice strange smells or sounds, or any other unexpected sensory illusion, turn off the circuit immediately. Consult your physician before attempting to use the project again.

Don't let these warnings scare you off from trying this project. Only a very tiny percentage of the population is at risk, but for those few people, the risk can be serious. The odds are very good that you will never suffer any ill effects and will find the relaxing effects of the alpha goggles quite beneficial.

A fairly good test of likely problems is a strobe light like those used in many discotheques. If strobe lights make you ill or dizzy, you probably should not try the alpha goggles. If strobe lights have never bothered you, then it is very likely that you won't experience any problems with this project. But let me repeat that epileptics should not even consider experimenting with this type of project.

❖3 Biofeedback monitors

BIOFEEDBACK IS PROBABLY THE MOST "RESPECTABLE" OF THE concepts covered in this book. The basic principles are fairly well-accepted in the general scientific community. In New Age circles, biofeedback is just used a little more extensively and with somewhat wider application. Serious work in the biofeedback area has been going on since the 1960s. Some of the preliminary research dates back to the 1930s or even earlier.

A fairly typical attitude about biofeedback was expressed by noted psychologist, Dr. Barbara Brown. She said, "Probably no discovery in medicine or psychology compares in breadth of application or in scope of implications to the biofeedback phenomenon." Some other experts might feel this is overstating the case a bit, but I doubt that anyone will deny that it is an important and fruitful area for research.

In many ways, biofeedback is closely interrelated with meditation and hypnosis (see chapter 2). In fact, the biofeedback monitors presented as projects in this chapter can be used as aids to induce meditative or self-hypnotic states.

Biofeedback theory

Just what is biofeedback all about? We know that the conscious mind uses only a small portion of the human brain's full potential. Many bodily functions are normally on "automatic pilot." We breathe and digest food and keep our hearts beating without consciously willing these activities or even giving them any conscious thought at all. Unconscious portions of the brain keep these internal mechanisms running so we can survive.

Some of these automatic bodily functions can be temporarily overridden by the conscious mind. For example, we don't usually have to think about breathing, but if we choose, we can change our breathing pattern. We can take deeper or shallower breathes. We can breathe through our noses or through our mouths. We can even hold our breath for short periods.

Other automatic functions appear to be outside the reach and control of the conscious mind. Most of us cannot will our heartbeat to speed up or slow down. We can't deliberately change our perspiration rate. Biofeedback allows us to bring such "fully automated" functions at least partially under conscious control. Laboratory experiments with biofeedback techniques suggest that there might be some credibility to the stories about yogis or fakirs who can slow down their bodily functions enough to appear dead and then revive themselves.

Actually, we can all indirectly influence these automated bodily functions to some extent through our emotions. We do it all the time, generally without fully realizing just what we are doing. Try a simple little experiment to demonstrate this for yourself right now. It will just take a minute or two and no equipment or preparation is required.

Close your eyes and imagine something scary. Imagine it is real, so you start to feel a little scared. As soon as you feel the emotion of fear, extra adrenaline is released in the body, your muscles tense up—ready for fight or flight—your heartbeat speeds up, and your perspiration rate increases. You have used your conscious mind to alter these "automatic" functions, albeit in a rather crude and clumsy manner. Biofeedback permits more direct and sophisticated control, usually with fewer undesired side effects. You can't use fear to increase your heart rate without a burst of adrenaline and a similar increase in the perspiration rate. With biofeedback, more selective control is possible.

The concept of feedback shouldn't be totally unfamiliar or difficult to understand for anyone involved with electronics. The name really says it all—biofeedback is simply biological feedback.

In any feedback system, some of the output (result of the system) is used to control the operation of the system itself. This is illustrated in the block diagram of a simple feedback system shown in Fig. 3-1. Feedback systems are extremely common in electronics. Feedback of some sort is essential for almost any oscillator circuit and for most amplifier circuits, and for many,

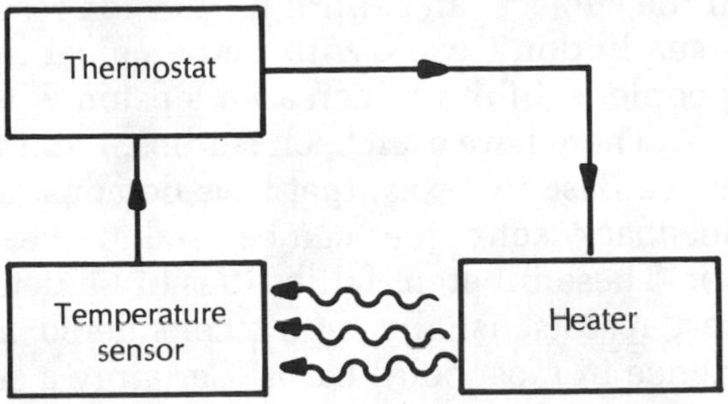

Fig. 3-1 A typical simple feedback system.

many other types of circuits as well. Surely anyone who works with electronics has at least some familiarity with the concept of feedback. Biofeedback works in the same way, except it involves biological, rather than electronic functions, hence the prefix "bio-."

A major reason why we normally can't consciously control these "automatic" bodily functions is that we have no practical way to directly observe them; that is, there is no feedback path. A biofeedback monitor gives us a sensory clue to the operation of the function we are trying to control; that is, a feedback path. The most commonly used type of sensory signal is audio. The volume or pitch of a tone is controlled by the function being monitored. The subject does not so much try to slow down his heartbeat, for example, as to lower the pitch of the tone sounded by the biofeedback monitor. This approach is surprisingly effective.

Why would anyone want to bother with biofeedback? Scientists can gain quite a bit of information about how the body functions by using such tools. Biofeedback monitors can also be used to reduce stress and counteract the undesired effects of various emotions. Another application of biofeedback is the lie detector. A so-called lie detector measures the galvanic skin response (GSR) of the subject being tested. For our purposes, we can simplify this somewhat and think of GSR as simple skin resistance. It's actually a little more complicated than that, but we don't need to be concerned with subtle differences.

When a person becomes tense, his perspiration rate increases, which lowers his skin resistance. The idea behind a lie detector is that for most people, the act of telling a lie causes some level of tension that can be detected through the changes in the GSR. A lie detector might be more accurately called a tension detector or even a fear detector. It is possible to fool a lie detector if the subject can control his tension response. Also lie detectors usually don't work with pathological liars who, unlike normal people, exhibit no increased tension when lying.

There have even been studies, including one at Lackland Air Force Base in Texas, that have demonstrated that training in biofeedback techniques can be used to ''beat'' or ''fool'' a lie detector. These inherent fallibilities of lie detectors explain why such test results usually aren't considered as admissible legal evidence in most court cases. Certainly a conviction should not be based solely on the results of a lie detector test.

In this chapter, several biofeedback monitor projects are presented. These circuits are designed to monitor some (normally automatic) bodily function and generate an appropriate sensory response—either audio or visual in nature.

Simple skin resistance biofeedback monitor

A very simple but effective biofeedback monitor can be built around a slightly modified astable multivibrator circuit, as shown in Fig. 3-2. This circuit is built around a 7555 CMOS timer IC. You can substitute a standard 555 timer chip if you prefer. The two chips are 100% compatible and no changes in the circuitry are required to accommodate the substitution. The complete parts list for this project is given in Table 3-1. You might want to experiment with some alternate component values. Nothing is particularly critical here.

The standard frequency formula for a 555 astable multivibrator is

$$F = \frac{1.44}{((R_a + 2R_b)C)}$$

where C is capacitor C1, R_a is resistor R1, and R_b is the series combination of resistor R2 and an unknown resistor connected between plates 1 and 2. Resistor R2 is included to prevent resistance R_b from dropping to 0 Ω if an accidental short circuit ever forms across the plates.

In operation, one finger is placed on each plate. Generally it is most convenient to place your index finger on plate 1 and the middle finger (of the same hand) on plate 2, but this is not absolutely necessary. Any two fingers will do. In effect, the skin resistance of the subject completes the astable multivibrator circuit.

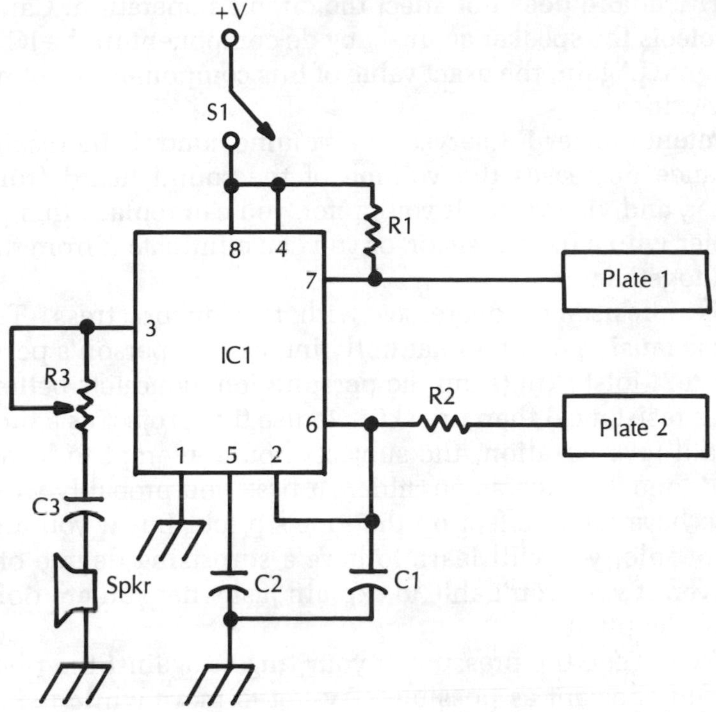

Fig. 3-2 *Simple skin resistance biofeedback monitor project.*

Table 3-1 Parts list for the simple skin resistance biofeedback monitor project of Fig. 3-2.

IC1	7555 timer (or 555)
C1	0.022-μF capacitor
C2	0.01-μF capacitor
C3	0.1-μF capacitor
R1	2.2-kΩ, 0.25-W, 5% resistor
R2	1-kΩ, 0.25-W, 5% resistor
R3	500-Ω potentiometer (volume)

The exact value of the subject's skin resistance will determine the output frequency. When the skin resistance increases, the output frequency decreases and a lower pitch is heard from the speaker. Similarly, when the skin resistance drops for any reason, the output frequency increases and the speaker emits a higher-pitched tone.

Capacitor C2 ensures the stability of the tier chip. It might not be essential in all cases, but it is cheap insurance against potential stability problems. The exact value of this capacitor is not critical and does not affect the circuit's operation. Capacitor C3 protects the speaker against any dc component in the IC's output signal. Again, the exact value of this component is not particularly critical.

Potentiometer R3 serves as a volume control. Increasing its resistance decreases the volume of the sound heard from the speaker, and vice versa. If you prefer, you can replace this potentiometer with a fixed resistor, or you can eliminate it from the circuit altogether.

Skin resistance decreases with tension or stress. This is because tension or stress naturally increases a person's perspiration rate. Moist skin (from the perspiration) conducts better (has a lower resistance) than dry skin. To use this project as a biofeedback aid for relaxation, the subject should attempt to lower the output tone as much as possible. At first, you probably won't be able to have much effect on the tone's pitch. But if you are like most people, you will learn to have a surprising degree of control, even if you aren't able to explain just what you are doing to control the pitch.

Try to keep the pressure of your fingers against the plates as even and constant as possible. Try not to move while using the biofeedback monitor because this will throw off the resistance and give a false output. Perhaps the best way to avoid such problems is to use small metallic discs for the contact plates. Attach these discs to strips of fabric (or some similar material) that can be wrapped around the fingers. Velcro or some other fastener can be used to hold the straps in place. A typical example of this approach is illustrated in Fig. 3-3.

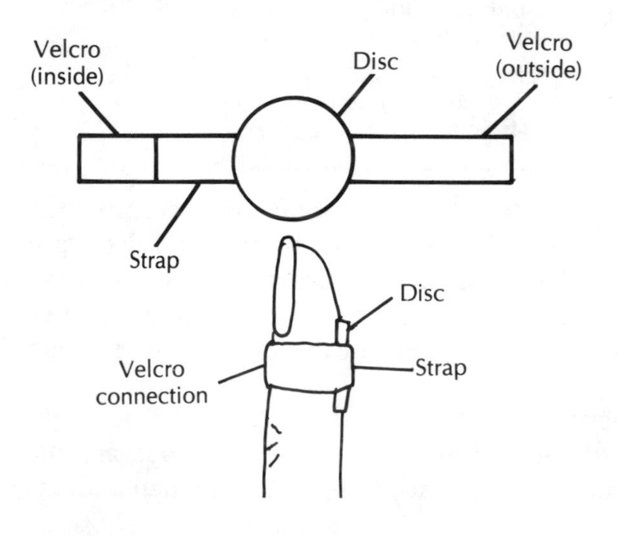

Fig. 3-3 *Fingertip discs held in place by straps with velcro will give more reliable results from a biofeedback monitor.*

Important! Because the subject's body is used as part of the circuit and is electrically connected to the circuit, it is vital that you *use battery power only* with this project. Never use an ac power source or adapter. The potential danger if there is a short circuit or other problem with the adapter is too great. It is not worth the risk.

This biofeedback monitor project can also be used as a simple lie detector, as you might have guessed from the introductory discussion at the beginning of this chapter. When a person tells a lie, his perspiration increases, his skin resistance decreases, and the pitch of the tone heard from the speaker goes up. Be aware that as a lie detector this project is extremely crude and not at all precise in its response. It should not be used for any application more serious than a casual party game. (But it can be a lot of fun when used for such purposes.) Don't take any lie detector results from this project too seriously. Many, many factors could "confuse" or "fool" such a simple lie detector circuit.

Computerized biofeedback monitor

For more sophisiticated analysis, along with storage and comparison of readings, you might want to try a computerized system. A very simple computerized biofeedback monitor interface circuit is illustrated in Fig. 3-4. A suitable parts list for this project appears in Table 3-2.

The interface to the computer is quite simple and generalized. It can be easily adapted to virtually any computer that permits external input and output connections. Only one input line and one output line are required, along with the common (ground) connection. This is about as simple as a computer interface can get. With some computers, some sort of buffering might be required to shape the signals for that particular machine's input and output requirements.

The circuit is a monostable multivibrator built around a 7555 CMOS timer or a 555 standard timer IC. These two chips are pin-for-pin compatible and either one will work without modification of the circuitry. As in the preceding project, two plates are used to test the subject's skin resistance as a timing "resistor" in the circuit. Let me repeat the warning from the previous project—because the subject's body is used as part of the circuit and is electrically connected to the circuit, it is vital that you

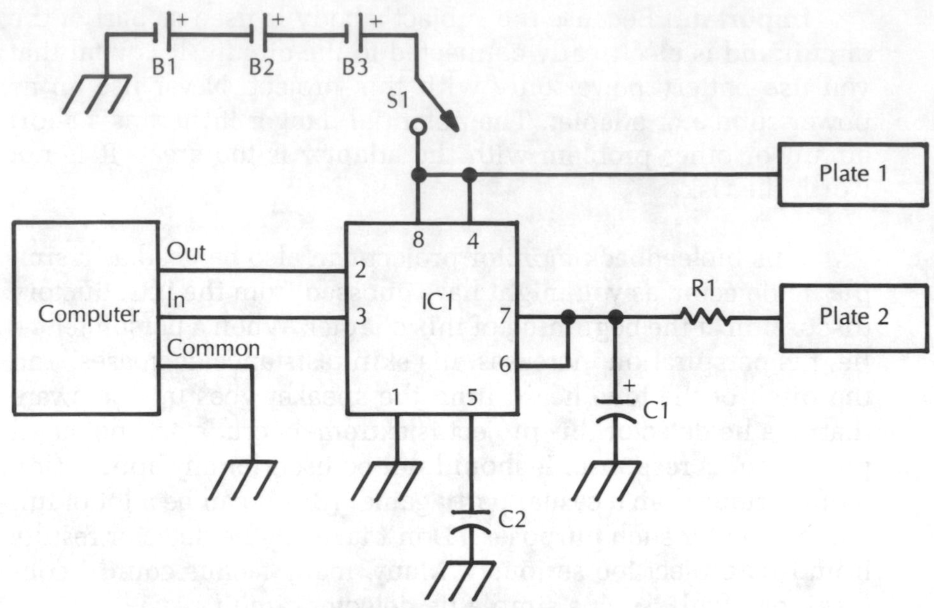

Fig. 3-4 Simple computerized biofeedback monitor project.

Table 3-2 Parts list for the simple computerized biofeedback monitor project of Fig. 3-4.

IC1	7555 timer (or 555)
C1	10-μF, 35-V electrolytic capacitor
C2	0.01-μF capacitor
R1	1-kΩ, 0.25-W, 5% resistor
B1, B2, B3	1.5-V battery
S1	SPST switch

USE BATTERY POWER ONLY with this project. Never use an ac power source or adapter. The potential danger if there is a short circuit or other problem with the adapter is too great. It is not worth the risk.

Using three 1.5-V penlight (AA) batteries will give a supply voltage of 4.5 V. The high level of the output voltage will be just slightly under the supply voltage. Most computers are designed to accept high levels up to 5 V, so be very careful about increasing the supply voltage to this project.

The subject's skin resistance (determined by his perspiration rate, which is linked to his stress level) determines the length of the output pulse when the circuit is triggered. An output bit from the computer is used to trigger the timer to take a reading. The output from the timer is fed into the computer as an input bit. The computer's programming should be set up to measure how long this input pulse is high. This time will be directly proportional to the subject's skin resistance at the time the measurement is taken.

The exact programming depends on the particular computer used and exactly what you want to use the project for. Therefore, I will not include any actual programs here. Because you only have to deal with a single input bit and a single output bit, it will not be difficult to write your own customized program.

Generally you'll get the best results by averaging together several closely spaced measurements. For example, the program can be written to send out an output bit (to trigger the timer) 2 seconds after the end of the last detected input pulse. Measurements can be stored in an updatable array of some sort. The display can be the average of the last five readings and some indication of how this average has changed over time, indicating an increase or decrease in stress. A large, easy-to-see graphics display is generally more effective than simply printing out the raw numbers. Many computers permit you to augment the display with color and sound, which can be very effective.

As with the preceding biofeedback monitor project, when using this device, try to keep the pressure of your fingers against the plates as even and consistent as possible. Try not to move while using the biofeedback monitor, because this will throw off the resistance and give a false output.

Perhaps the best way to avoid such problems is to use small metallic discs for the contact plates. Attach these discs to strips of fabric (or some similar material) that can be wrapped around the fingers. Velcro or some other fastener can be used to hold the straps in place. Refer to Fig. 3-3 for a suggested approach to making suitable fingertip bands for your biofeedback monitor project.

Biofeedback temperature monitor

Skin resistance (perspiration rate) is certainly not the only bodily function that can be used in biofeedback applications. This next project will help you learn to partially control your body temperature. To a limited extent, when it is cold, you can warm yourself up a little. On hot days, you can cool yourself off a bit. You might be able to bring down a fever. (Remember, you would only be affecting a symptom here—the underlying medical problem would remain. None of these projects should ever be used as a substitute for a qualified doctor or prescribed medication.)

Actually, what you are doing through biofeedback is adjusting your blood flow, which alters your subjective temperature—how hot or cold you feel. Don't expect miracles. The effect is a little like a freezing person drinking alcohol and feeling quite warm as he freezes to death. All the alcohol does is bring the blood flow closer to the surface of the skin, making the person feel warmer without altering the true problem—the dangerously cold environmental conditions. Still conscious control over your subjective body temperature can be very useful within the natural limits of the technique. You can learn to feel comfortable under a wider range of conditions. If nothing else, this will permit you to save a little energy on your heating and air conditioning bills. Don't expect to make major changes in your body temperature. At most, you'll probably be able to change it by a few degrees. But this can be enough to make a noticeable difference on the subjective level.

In some cases, the improved circulation might make migraine headaches less painful. Remember, once again, we are only treating a symptom to ease discomfort; it is not a cure or remedy. The underlying medical problem remains and should be treated. Biofeedback control acts like a self-created natural aspirin or painkiller for migraine sufferers. It temporarily relieves certain symptoms until more permanent help can be obtained. Of course, this in itself is valuable and useful, providing you don't forget its limitations. Obscuring the symptoms does not remove the problem.

Like most other biofeedback techniques, temperature biofeedback is also useful for aiding relaxation and meditation and for reducing stress. As far back as the late 1930s, researchers B. Mittelmann and H. G. Wolff discovered that material with strong emotional associations can lower the temperature of a subject's hands. By 1976, a study at Duke University demonstrated that certain emotional states (such as stress, anxiety, and excitement) have physiological effects, including a constriction of the blood vessels in the hands and feet. Because of this constriction, blood flow is reduced, resulting in a lower skin temperature.

It is therefore not unreasonable to expect these effects to work in the opposite direction (as is the case with many other bodily functions). That is, if we can increase the skin temperature (expand the blood vessels and increase the blood flow), any emotional turmoil should be soothed—we find ourselves growing relaxed.

To accomplish this conscious control of body temperature, you need some way to dynamically monitor your body temperature. The circuit to do this is shown in Fig. 3-5. A suitable parts list for this project is given in Table 3-3.

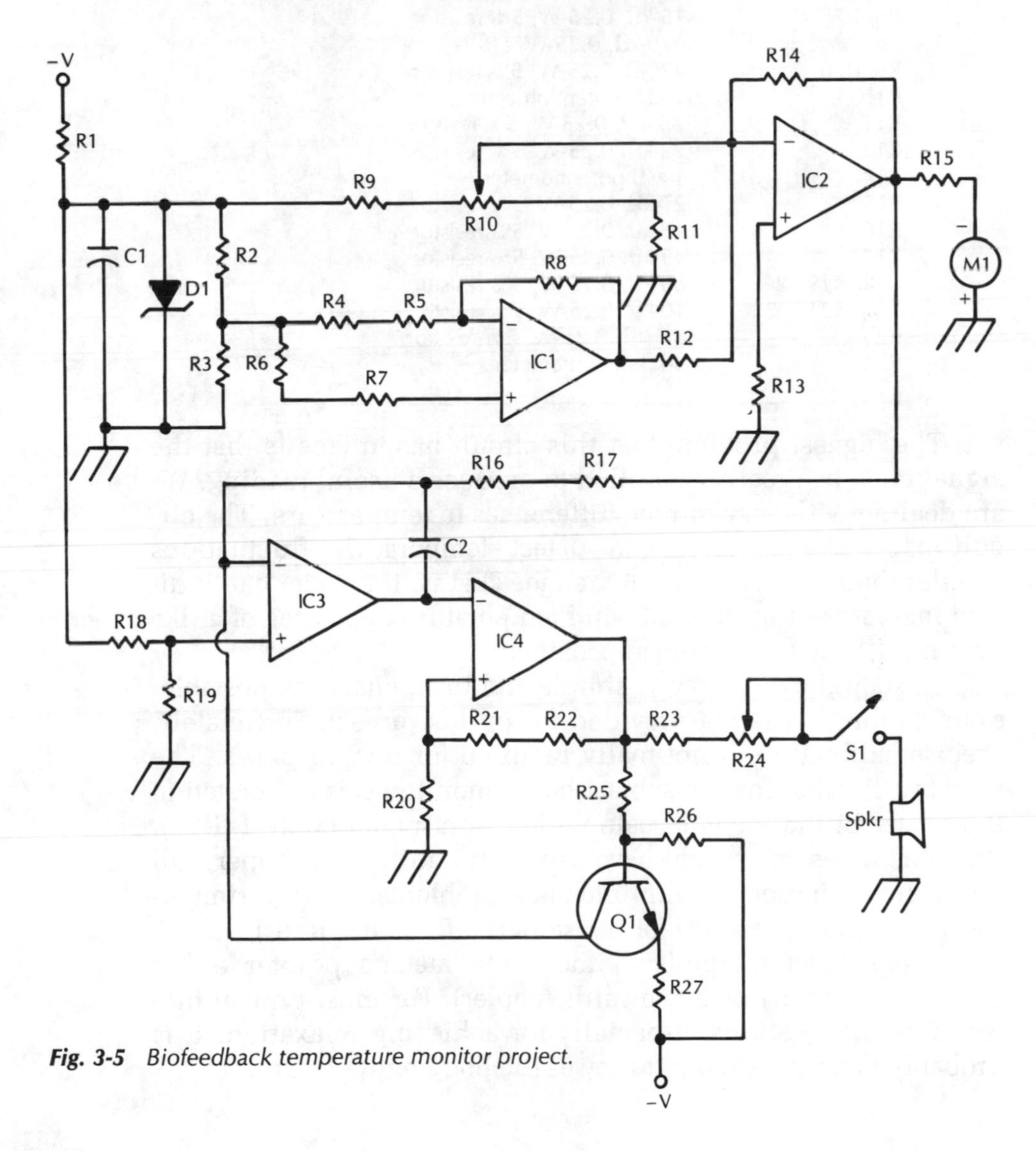

Fig. 3-5 Biofeedback temperature monitor project.

Table 3-3 Parts list for the biofeedback temperature monitor project of Fig. 3-5.

IC1 – IC4	Op amp (LM324 quad op amp or similar)
Q1	npn transistor (Radio Shack RS1617, GE-10, HEP55, SK3020, or similar)
D1	5.6-V zener diode
C1	0.1-μF capacitor
C2	0.47-μF capacitor
R1, R3, R27	1-kΩ, 0.25-W, 5% resistor
R2, R25	4.7-kΩ, 0.25-W, 5% resistor
R4, R6, R12	47-kΩ, 0.25-W, 5% resistor
R5, R7	15-kΩ, 0.25-W, 5% resistor
R8	470-kΩ, 0.25-W, 5% resistor
R9, R26	470-Ω, 0.25-W, 5% resistor
R10	10-kΩ potentiometer
R11	2.2-kΩ, 0.25-W, 5% resistor
R13	22-kΩ, 0.25-W, 5% resistor
R14	1-MΩ potentiometer
R15	27-kΩ, 0.25-W, 5% resistor
R16	18-kΩ, 0.25-W, 5% resistor
R17	39-kΩ, 0.25-W, 5% resistor
R18, R19	68-kΩ, 0.25-W, 5% resistor
R20, R21, R22	10-kΩ, 0.25-W, 5% resistor
R23	3.3-kΩ, 0.25-W, 5% resistor
R24	5-kΩ potentiometer

The biggest problem that this circuit has to face is that the signal must be greatly amplified to produce a useful reading. We are dealing with very minute differences in temperature. The circuit as presented here can detect temperature fluctuations smaller than 0.1°F over a 2.5° range. That is, the meter can indicate increases of up to 1.25° and temperature decreases of a like amount. These figures are approximate.

To keep the circuitry as simple and inexpensive as possible, exact calibration is not provided for in this project. Fortunately, precise calibration is not really required for our purposes. The base (null) value for the subject being monitored is calibrated to the center of the meter's scale with a front-panel control (R10). (In some cases you might have difficulty setting the proper null point. If you happen to run into such problems, try experimenting with different values for resistors R2, R18, and R19.)

The subject attempts to move the meter's pointer either upwards (warmer) or downwards (cooler). For most typical biofeedback applications, especially for achieving relaxation, it is probably better to attempt to lower the body temperature.

This biofeedback monitor project also has a switchable audio output, which many people find easier to use for biofeedback than watching a meter. The subject attempts to raise or lower the pitch of the tone from the speaker. Potentiometer R24 is a volume control for the speaker. This potentiometer can be replaced with a fixed resistor of a suitable value if you prefer.

Four op amps are used in this project. They don't need to be high-grade, precision devices. Standard, garden-variety 741-type op amps will do just fine. You can use four separate single op amp ICs, two dual op amp ICs, or a single quad op amp IC, depending on what you happen to have available.

Like most circuits using 741-type op amps, a dual-polarity power supply is required for this project. V+ should be +9 V and V− should be −9 V, both referenced to the circuit ground. A pair of 9-V "transistor" batteries will work fine. In Fig. 3-5, power supply connections are not shown to the op amps themselves to avoid cluttering the schematic, making it unnecessarily difficult to read. It is essential, of course, to connect both V+ and V− to each and every IC in the circuit. Notice that there are two points in the schematic that are to be connected to the negative power supply (V−). The positive power supply (V+) is not used in this circuit, except by the op amps.

The thermistor (R28) is the actual temperature probe in this project. It should be mounted at the end of an insulated cable. The subject can hold the thermistor probe gently but firmly in a fist, or the thermistor can be taped (with a band-aid or something similar) to the subject. The forehead is a good spot. The wrist is also a good choice. These seem to be relatively easy places for most people to learn to control their temperature. The advantage of taping the probe in place, of course, is that this minimizes confusing misreadings due to movement by the subject.

Capacitive biofeedback monitor

Earlier in this chapter we looked at some biofeedback monitors that detected changes in the subject's skin resistance. The human body also exhibits capacitance, which offers us yet another approach to biofeedback.

The circuit diagram for our capacitive biofeedback monitor project is shown in Fig. 3-6. A suitable parts list for this project appears in Table 3-4. For this project you will need to make your

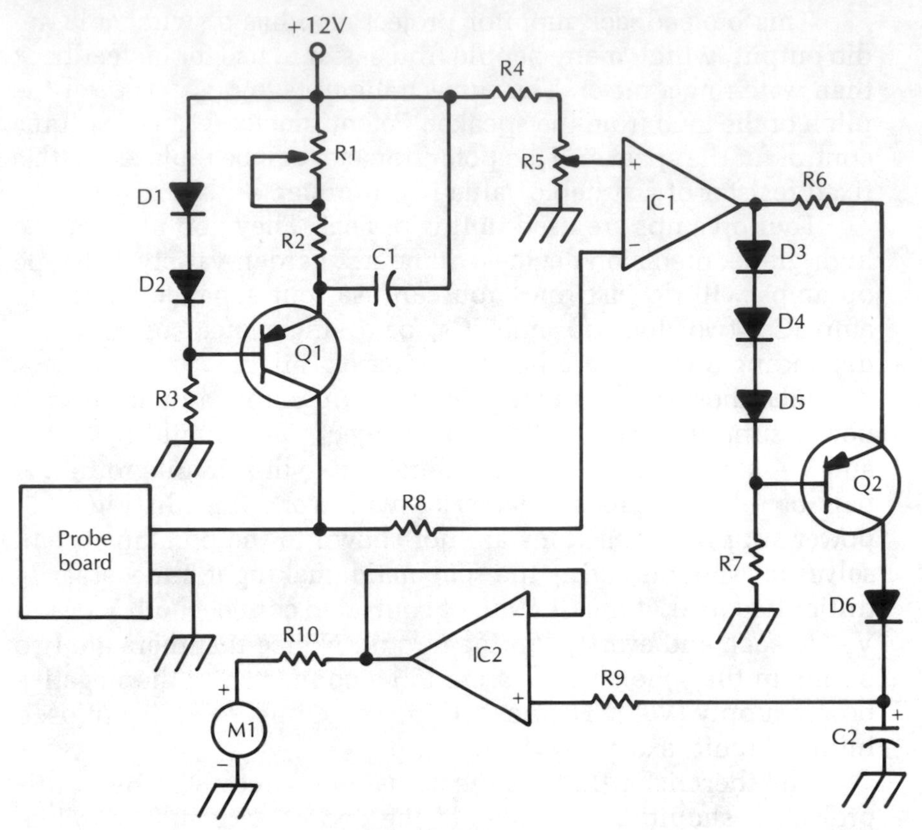

Fig. 3-6 Capacitive biofeedback monitor project.

Table 3-4 Parts list for the capacitive biofeedback monitor of Fig. 3-6.

IC1, IC2	Op amp
Q1, Q2	pnp transistor (Radio Shack RS2023, GE-22, HEP716, SK3025, or similar)
D1 – D6	Diode (1N4148 or similar)
C1	0.01-μF capacitor
C2	25-μF, 35-V electrolytic capacitor
R1	1-MΩ potentiometer
R2	470-Ω, 0.25-W, 5% resistor
R3, R7	22-kΩ, 0.25-W, 5% resistor
R4	100-kΩ, 0.25-W, 5% resistor
R5	5-kΩ potentiometer
R6	33-kΩ, 0.25-W, 5% resistor
R8	12-kΩ, 0.25-W, 5% resistor
R9	10-kΩ, 0.25-W, 5% resistor
R10	1-kΩ, 0.25-W, 5% resistor
M1	1-mA milliammeter

own capacitance board. This is not a very difficult or complicated job. It is simply a matter of etching an interlaced pattern on a piece of copper-clad board. A typical pattern is shown in Fig. 3-7. The exact arrangement and dimensions of this pattern are not overly critical. The important thing is that the interlaced legs should be relatively thin and closely spaced. Notice that there are two separate electrical paths that do not go anywhere and do not connect. These serve as the two plates of the capacitor.

The complete capacitance board should be fairly small. Two or three inches square is sufficient in most cases. It might be helpful in some applications to affix the capacitance board to a larger base. Connect the base to the back of the board only—do not connect it on the side with the copper interlace pattern.

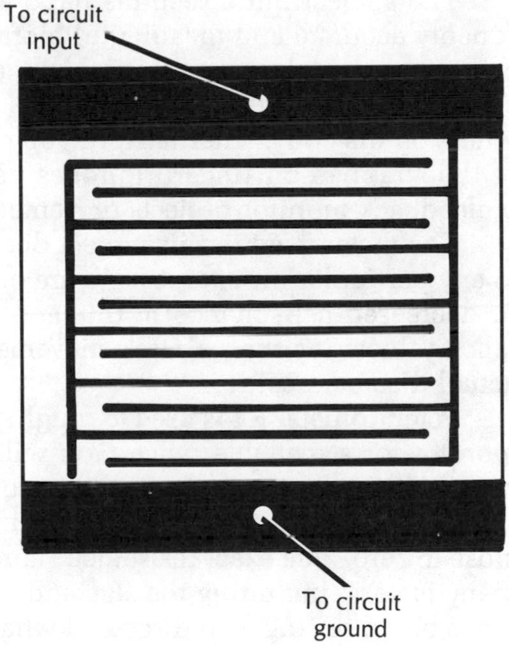

Fig. 3-7 A PC board with an interlaced pattern serves as the probe in the capacitive biofeedback monitor project of Fig. 3-6.

The subject's hand is placed over the interlaced pattern, as illustrated in Fig. 3-8. The subject's hand (or other body part) serves as the dielectric in a pseudocapacitor. The etched legs on the PC board make up the plates of this pseudocapacitor. This hand capacitance effect is very sensitive to variations in the pressure applied. In fact, this type of board can be used as a pressure sensor. The capacitance board is so sensitive to fluctuations in the applied hand pressure that this gives us a handy, "quick and dirty" way to test out the circuitry. Deliberately try a light touch

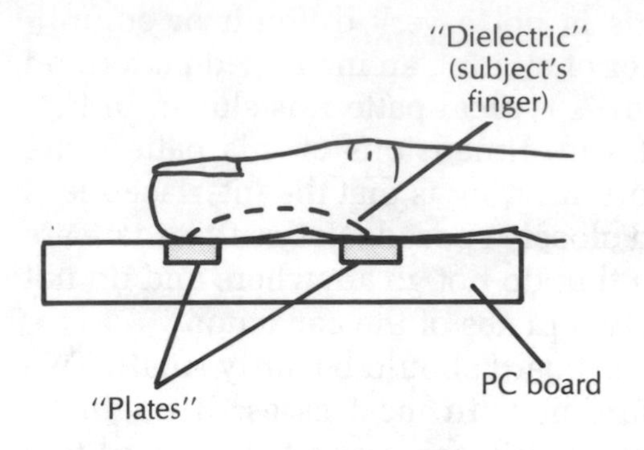

Fig. 3-8 The subject's hand acts like a dielectric between the plates of the capacitive hand probe board of Fig. 3-7.

then a heavier touch to determine if the pointer is moving noticeably across the meter's face.

The subject must keep his hand as still as possible for reasonably accurate and meaningful biofeedback readings. In some cases, it might be more convenient and practical to tape the capacitance board directly against the skin of the subject, somewhere on his body. Alternatively, you can use a modified version of the strapped-on fingertip probes used in the skin resistance biofeedback monitor projects presented earlier.

The meter used in this project does not have to be calibrated to any particular unit values. We are only interested in increased or decreased capacitance in this application, not exact capacitance values. It's the pointer's movement that counts and not its actual discrete position.

Potentiometer R1 is used to roughly calibrate the biofeedback monitor for a readable range. You will probably have to experiment with different settings of this control until you find a setting that does not peg the meter's pointer or leave it unreadably close to zero. The exact resistance value required will depend on many factors, including the size and exact pattern of the capacitance board or fingertip discs and what part of the body is being monitored. For example, if you are using fingertip probes, you will get vastly different capacitances when you use two fingers on the same hand or one finger from each hand.

Nothing is critical in this circuit because we aren't particularly interested in measuring exact, calibrated capacitance values. Any inaccuracies within the circuitry are likely to be irrelevant in this application. Almost any low-power pnp transistors will work for Q1 and Q2. Both transistors should be of the same type number, though. Similarly, almost any standard signal diode will work for D1 through D6.

You certainly do not need to use expensive, high-precision op amps in this circuit. Standard 741s will give more than adequate performance. Substituting a high-grade chip probably wouldn't make any noticeable difference in the operation of the project, but it could make a substantial difference in the project's cost. You can use two separate 741 chips, a dual 747, or a 1458 chip. If you are using this circuit as part of a larger system, it might make sense to use two sections of a 324 quad op amp IC.

You might need to experiment with alternate values for capacitor C2 to prevent transistor Q1 (which functions as a constant current source) from breaking into oscillation. This is not likely to be a problem, but it is possible. If you have oscillation problems with your project, the surest and simplest cure is to change the value of capacitor C2.

Basically this circuit works by charging up the pseudocapacitor (the probe) with a known current from the constant current source, Q1, then measuring the time it takes this pseudocapacitor to discharge through a fixed resistance. The meter's pointer indicates the measured discharge rate and, thus, the capacitance value.

Because the subject's body is electrically part of the circuit, it is absolutely vital to use battery power only with this project. Never use any sort of ac adapter to power such a biofeedback monitor. If there is a short circuit or some other defect, the subject could receive a nasty electrical shock. At best, this would be painful. It is likely to be dangerous or even fatal. Please, don't take chances and stupid risks—USE BATTERY POWER ONLY!

Alpha-wave biofeedback monitor

In the last chapter we looked at brain waves, particularly alpha waves, which are indicative of deep relaxation. Our next biofeedback project also uses alpha waves, so we will very briefly review the relevant material here.

There are a number of different types of brain waves, each with its own frequency. These frequencies vary somewhat from person to person, but usually fall within certain ranges. The main classifications of human brain waves are as follows:

- Delta—0.5 – 3.5 Hz (sleep, illness)
- Theta—3.5 – 7 Hz (sleep, deep trance, problem solving)
- Alpha—7 – 14 Hz (relaxation)
- Beta—14 – 30 Hz (ordinary waking consciousness)

To ease stress and maximize relaxation, we want to increase the subject's alpha-wave level. Biofeedback can be used to teach the subject how to do this. Of course, that is precisely the purpose of this project. Unlike the alpha goggles project presented in chapter 2, this project should not present any risk to epileptics because there is no external alpha frequency for their brains to attempt to lock onto. Still, if you are an epileptic or have any related problems or symptoms, it would probably be a good idea to check with your physician first, just to be on the safe side.

The schematic diagram for the alpha-wave biofeedback monitor is shown in Fig. 3-9. A suitable parts list for this project appears in Table 3-5. Standard 741 op amps will work for IC1 and IC2, but better performance might be achieved with high-grade, low-noise op amps. Notice that the power supply connections to the op amps are not shown in the schematic diagram; this is to avoid clutter. Remember, an op amp IC must always have the proper power supply voltages applied to it to function properly and without damage.

A CMOS 7555 timer is recommended for IC4, but a standard 555 timer chip can be substituted. IC3 is an optoisolator. It ensures that there is no electrical connection between the input/amplification stage and the audio output stage. The output device in this optoisolator is an npn transistor. The exact specifications of this component are not critical and almost any optoisolator with an npn phototransistor output should work fine. (pnp phototransistors are actually rather rare.)

Transistors Q1 and Q2 are N-channel FETs (field effect transistors). They should be as closely matched as possible. If you can find a dual FET, such as the 2N5524, it would definitely be your best choice, but such components can be expensive and rather difficult to find. A dual FET, of course, is just two complete (and identical) FETs enclosed in a single, compact housing. The other transistors in this circuit are not critical and can be substituted with similar devices, depending on availability. Transistors Q5 and Q6 should be of the same type number; that is, use the same substitute for both of these components.

Three power supply voltages are required for this project— +15 V, +9 V, −9 V. Use battery power only for the +9-V and −9-V supplies. A pair of transistor batteries will do just fine. Remember that the subject is in direct electrical contact with the input circuit. If an ac power supply is used there will be the very real possibility of a short circuit or some other defect causing a

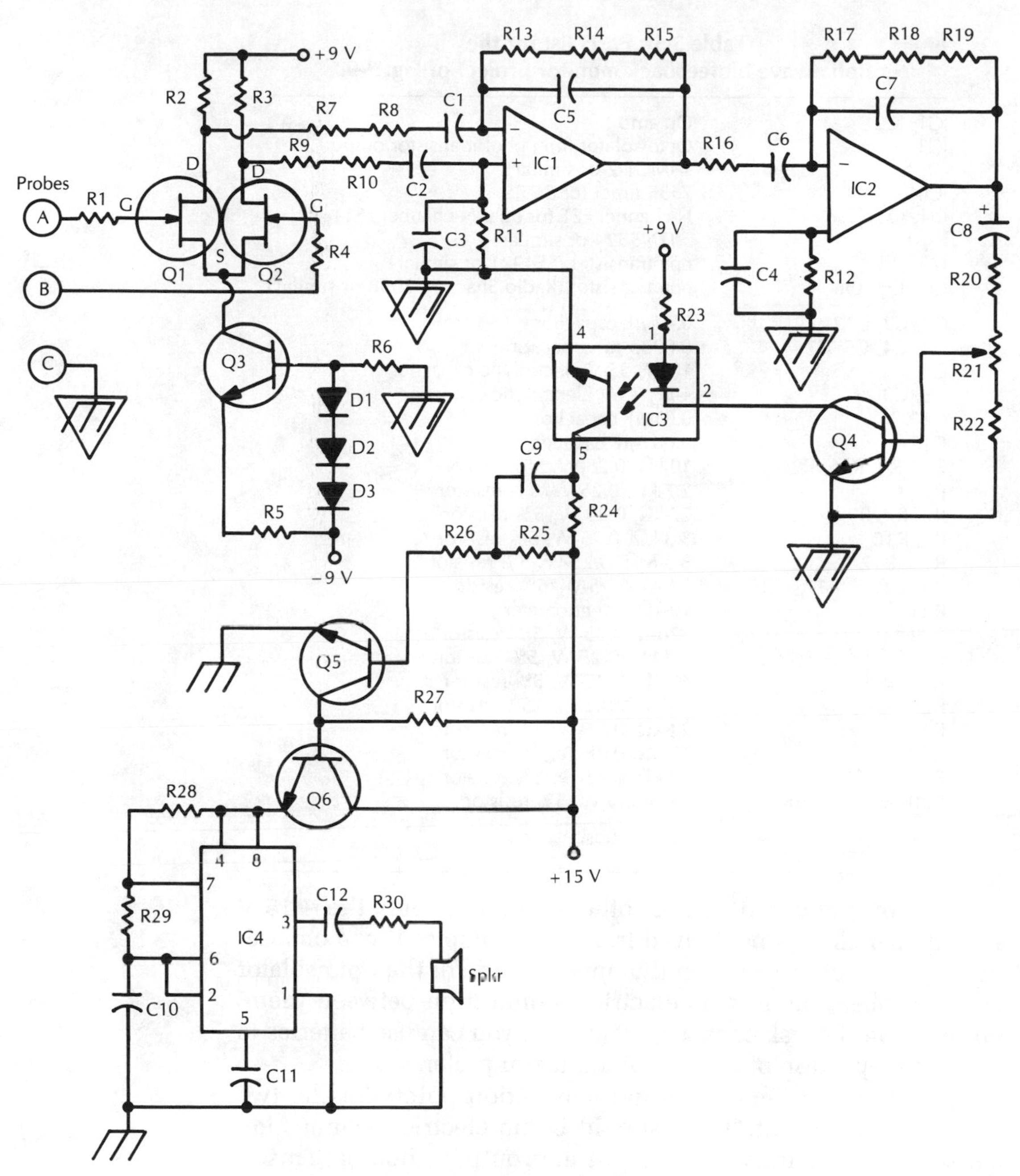

Fig. 3-9 Alpha-wave biofeedback monitor project.

painful and potentially dangerous (perhaps fatal) electrical shock. Don't take foolish chances! USE BATTERY POWER ONLY for the input circuit.

Table 3-5 Parts list for the alpha-wave biofeedback monitor project of Fig. 3-9.

Part	Description
IC1, IC2	Op amp
IC3	Optoisolator (npn phototransistor output) (MCT-2 or similar)
IC4	7555 timer (or 555)
Q1, Q2	N-channel FET (or dual N-channel FET) (2N5524 or similar)
Q3	npn transistor (SE4021 or similar)
Q4, Q5, Q6	npn transistor (Radio Shack RS2058 or similar)
C1, C2, C12	0.47-μF capacitor
C3, C4, C5, C7	0.0033-μF capacitor
C6	10-μF, 35-V electrolytic capacitor
C8, C9	1-μF, 35-V electrolytic capacitor
C10	0.01-μF capacitor
C11	0.001-μF capacitor
R1 – R4, R16, R20	10-kΩ, 0.25-W, 5% resistor
R5	2.7-kΩ, 0.25-W, 5% resistor
R6, R7, R9	27-kΩ, 0.25-W, 5% resistor
R8, R10	3.3-kΩ, 0.25-W, 5% resistor
R11, R12	3.3-MΩ, 0.25-W, 5% resistor
R13 – R15, R17 – R19	1-MΩ, 0.25-W, 5% resistor
R21	10-kΩ potentiometer
R22	470-Ω, 0.25-W, 5% resistor
R23	4.7-kΩ, 0.25-W, 5% resistor
R24, R29	47-kΩ, 0.25-W, 5% resistor
R25	100-kΩ, 0.25-W, 5% resistor
R26	22-kΩ, 0.25-W, 5% resistor
R27	15-kΩ, 0.25-W, 5% resistor
R28	12-kΩ, 0.25-W, 5% resistor
R30	33-Ω, 0.5-W, 5% resistor
SPKR	Small loudspeaker

The output circuit, on the other hand, uses only the + 15-V supply, which may be derived from an ac adapter if you choose. Because this circuitry is on the opposite side of the optoisolator as the probes, there is no electrical connection between them, eliminating the risk of shock. Of course, you can use batteries to power this portion of the circuit too if you prefer.

Use totally different ground connection points for the two halves of the circuit. There should be no electrical connection between the grounds of the input and output circuits. This is indicated in the schematic by the triangles around the ground symbols in the input circuit. This indicates a totally different ground than the "normal" ground used for the + 15-V power supply. In other words, you actually have two distinct and separate circuits in this project, and there should be absolutely no electrical connection between the two except through the opto-isolator (IC3).

The probes are placed against the subject's forehead and scalp. The best way to do this is to affix metallic discs inside a snug-fitting headband. You might need to experiment a little to find the best possible placement for the probes. Try to avoid getting any of the subject's hair between the discs and the subject's skin. Probe A should be placed over the right frontal lobe, and probe B should be positioned at the rear of the skull, opposite the left occipital lobe of the brain. The ground probe (C) should be placed in a neutral or in-between position.

The input section of the circuit detects and amplifies the difference between the electrical signals being picked up by probes A and B. To minimize interference, it is a good idea to twist the input wires between the probes and the circuitry. Shielded cable is also advisable. Remember, we are dealing with very tiny electrical signals. It doesn't take much interference to drown out the desired signals entirely. The input circuit is designed to reject common-node voltages as much as possible. Most interference signals will appear equally on both input lines, so they can be canceled out unless the problem is extremely severe. Remember also that the alpha-wave measurement performed by this project is quite crude. Do not mistake this (or any similar) project for a piece of medical or therapeutic equipment.

The output of this circuit is a series of beeps emitting from the speaker. A beep will be heard for each detected brain wave pulse. Use the biofeedback monitor to attempt to slow the beeps as much as possible. This will seem difficult or even impossible at first. Just relax and be patient. Eventually, through biofeedback, such control will seem as natural as breathing. Once you are fully trained in the biofeedback technique, you will be able to emphasize your alpha waves (and enjoy the benefits of the resulting relaxation) at any time, even without the biofeedback monitor.

If the subject is very stressed, a harsh continuous tone might be heard from the speaker instead of discrete beeps. Separate beeps are still being sounded, but they are coming too fast for the human ear to distinguish between them, so they blend together into a single tone with a lot of enharmonic sidebands. As the subject relaxes, the signal should slow down into a string of distinguishable beeps.

If you don't care for the tonal quality emitting from the

speaker, experiment with alternate values for capacitor C10 and resistors R28 and R29. IC4 and its associated components are just a standard astable multivibrator circuit. Power to this multivibrator is turned on and off by transistors Q5 and Q6 in step with the detected input (brain wave) pulses.

If the volume of the beeps is too high for comfort, increase the value of resistor R30. Similarly, reducing the resistance of this component will produce louder output beeps. The "best" resistance value will depend upon personal preference and the size and individual design characteristics of the particular speaker you use in your project.

Some final words about biofeedback

Some people have considerable difficulty getting the "hang" of biofeedback. They try too hard and don't accomplish anything at all. With some subjects, other meditation or hypnosis techniques might be needed at first. You can use the projects presented in chapter 2 in the early stages of biofeedback training, but such extra aids should not be necessary for long.

Of course, some people have a strong unconscious resistance against relaxation. They might need professional help before they can take full advantage of these biofeedback monitors. Such problems are rather rare, however. For most people, such resistance (if any) is only temporary and quickly dissolves.

Remember, biofeedback is *not* dangerous. Your brain will not be damaged or affected in any way—no more so than the difference between thinking about something very unpleasant and stressful and something very pleasant and relaxing. The biofeedback process is 100% natural. It's just a tool to help you notice subtle cues that are normally below the level of your conscious attention. You will not slip into an inescapable trance or anything like that. The very worst that could happen is that you might fall asleep. If this happens, just enjoy your nap. No harm done.

4 ESP testers

EXTRASENSORY PERCEPTION IS ONE NEW AGE CONCEPT THAT HAS gotten a lot of attention over the years. This concept (or more properly, group of concepts) is not solely the province of the New Age. After all, there was discussion about ESP long before the New Age movement got started. But then, many New Age concepts aren't really new, except perhaps in the ways they are grouped together philosophically. The term "ESP" covers a wide range of mental and psychic phenomena, mostly variations on the "mind over matter" concept. A person with ESP can do things with his mind beyond what is normally accepted as possible.

Types of ESP

Strictly speaking, ESP refers to phenomena relating to obtaining information without the use of the ordinary five senses (sight, hearing, smell, taste, and touch). That is, an extra sense of some sort is implied hence, "extrasensory perception." Typical examples include telepathy, or "mind reading," and clairvoyance.

Telepathy is a form of mind-to-mind contact or a form of communication on a purely mental, nonphysical level. One person can "hear" or otherwise sense another person's thoughts; or it can work in the opposite direction. Telepathy can supposedly be used to send or implant a thought in another person's mind.

"Clairvoyance" literally means "clear seeing." A person with clairvoyance can mentally "see" or otherwise sense something that is happening (has happened or will happen) at some distance, well beyond the limits of his ordinary senses. A variation on clairvoyance is foreknowledge, which is knowing what is going to happen before it does. A person with such clairvoyant abilities is often referred to as a "psychic." When a psychic gives a clairvoyant report on a specific individual it is called a "reading."

Other unusual (and presumably impossible according to traditional scientific thinking) mental powers are also commonly lumped under the heading "ESP," even if they aren't directly related to the senses or perception. An example of this is *telekinesis*, which is the ability to move or otherwise affect material objects solely through the power of the mind.

The claims made for various ESP phenomena certainly seem far-fetched, until you stop to remember that the ordinary person uses less than 10% of the brain's capabilities. What could that "extra" 90% do if it could be accessed?

A great many studies have been made into the ESP phenomenon over the last few decades. Thus far the laboratory evidence for ESP has been highly questionable, though often intriguing and suggestive. Clear-cut proof seems to be difficult to obtain in a controlled setting. Anecdotal evidence of ESP phenomena occurring naturally in the "real world" (outside the laboratory) abound, but these often cannot be conclusively confirmed (or denied). Hard scientific evidence has been quite elusive.

Some believers in ESP have suggested that nonbelieving scientific researchers are unconsciously using their own unadmitted ESP powers to influence and disguise the results of their laboratory tests. But such an "explanation" is hardly satisfying or convincing. Perhaps a more likely explanation is that natural ESP works best when there is strong emotional motivation involved. Most (though not all) anecdotes about ESP experiences suggest that the mysteriously obtained information was of some emotional importance to the receiver or someone close to him. Perhaps the cut-and-dried atmosphere of a scientific laboratory impedes that ESP forces so only muddled results can be obtained.

Of course, we still don't have enough solid information to rule out some other more "natural" explanation for the voluminous ESP-related anecdotal evidence that has accumulated over the years. It seems unlikely that all apparent cases of ESP are true hoaxes or delusions. There have been far too many inexplicable incidents where a hoax or a delusion doesn't seem to be a very satisfactory or convincing explanation. But that doesn't prove that ESP exists. There may be something else going on, and we simply don't have all the pieces to the puzzle yet.

Again, as with most of the concepts discussed in this book, greater study and research is needed before we can say with certainty that, "ESP is real," or "ESP is not real." To be dogmatic about either position at this time is rather unrealistic. Of course, you are entitled to believe whatever you want, but don't present your beliefs (positive or negative) as facts.

Testing ESP

In this chapter I will present several projects that you can use to perform your own experiments in ESP phenomena, particularly telepathy, foreknowledge, and clairvoyance. Don't place too much importance on any one test. There is almost certainly a large element of chance involved. With just a few tests, you can't tell if a given subject truly has ESP or has just made a few lucky guesses. To be meaningful, such experiments must be repeated over and over until you have obtained a very large body of data.

Let's assume you are performing a simple two-possibility test. The subject has to choose either possibility A or possibility B for each test. Even if the subject is simply guessing, there is a 50% chance the subject will be right on any given test. Therefore, if you test that subject 10 times and the subject gets 7 right, that isn't anything to get excited about. Yes, the subject got 70% right, but pure chance suggests the subject would get 50%, and there are so few samples that such results are well within the realm of chance. The laws of probability are not stretched by such results. On the other hand, if you test that subject 500 times and the subject comes out right 350 times (again, 70%) that is much more impressive (though still hardly conclusive). Being off from the probability norm (50%) by 150 is a lot less likely to be pure chance than being off by just 2 samples when only 10 tests were taken. But remember, it is still possible to be pure chance even with the larger set of samples, though this becomes increasingly less likely as the sample size is increased. Even with the largest imaginable sample size, it is theoretically possible (though very, very unlikely) for someone to guess 100% right through pure chance.

When accumulating your data, remember it would be just as significant if someone is wrong more often than pure chance probability would suggest. If a second subject takes 500 tests in our two-possibility test and gets only 137 right, this should call for some notice. Could they be using ESP " backwards" somehow? Could they be, on some unconscious level, fighting or disguising their ESP abilities? Or is it just an unlikely, but possible, result of random chance? Such questions are not easy to answer, and it is foolish and very unscientific to glibly jump to conclusions in either direction.

Two-choice ESP tester

The simplest ESP tests involve selecting one of two possible events. For scientific validity, these events must be randomly selected. If a person just made up a list of 1s and 2s, it is likely that an unconscious pattern will sneak in. But if an electronic circuit makes the "choices" they will be more truly random.

A simple random-choice circuit is shown in Fig. 4-1. It is the electronic equivalent of flipping a coin. It has two possible output conditions, one of which is randomly displayed each time the circuit is activated. There is no way (short of ESP) to accurately and consistently predict which output will turn up on any given test. A suitable parts list for this project is given in Table 4-1.

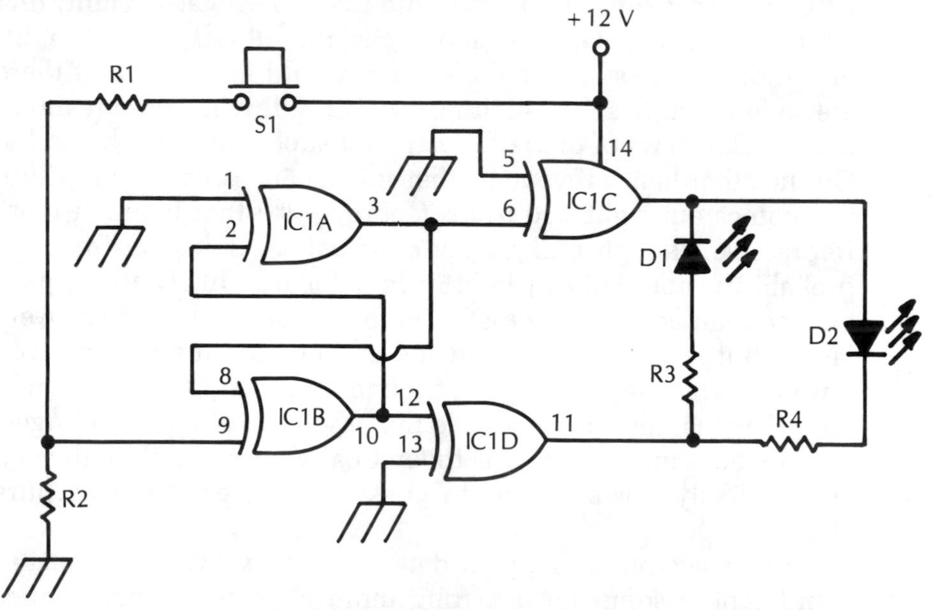

Fig. 4-1 Two-choice ESP tester project.

Table 4-1 Parts list for the two-choice ESP tester project of Fig. 4-1.

IC1	CD4070 quad X-OR gate
D1, D2	LED
R1	1-kΩ, 0.25-W, 5% resistor
R2	47-kΩ, 0.25-W, 5% resistor
R3, R4	330-Ω, 0.25-W, 5% resistor
S1	Normally open SPST push-button switch

There are two basic ways this project can be used as an ESP tester. To test clairvoyance or foreknowledge, the subject should try to predict beforehand which output will turn up the next time the circuit is activated. To test telepathy, a second person uses the circuit in a separate room with no direct contact with the subject being tested. This second person activates the circuit, then concentrates on the displayed output while the subject attempts to read his mind. In either case, a large number of tests are needed to have any scientific validity. Remember, because there are only two possibilities there is a 50% chance of randomly guessing correctly on any given test. Look for long-term statistical trends over many repeated tests.

To operate this project, first apply power to the circuit. While + 12 V is specified in the schematic, anything in the + 9-V to + 15-V range will work. Next, momentarily close push-button switch S1. Both LEDs (D1 and D2) will light up, and both will appear to be continuously lit as long as the switch is closed. Actually, they are alternately blinking on and off under the control of an oscillator formed by IC1. The LEDs are arranged for opposite polarities; when one is lit the other is dark. One and only one of the LEDs is lit at any given instant. But the oscillator frequency is so high that the human eye cannot detect the individual flashes of each LED. When the push button is released (the switch is opened), the oscillator latches onto its last output state. One of the LEDs remains lit, while the other goes dark. Because the oscillator is set so that the LEDs blink on and off many times each second (much, much faster than human reflexes), there is no way to consciously control the final output. The effect is a truly random output of the two possible states.

I recommend using different colors for the two LEDs—say, red for D1 and green for D2. This will make it easier to identify the output state. If, for example, the test subject says that the one on the right will be lit, there is the question of which way the observer is looking at the project. It is also easier to identify the two output states simply as "red" and "green." Even a child can use the system.

Of course, this circuit can be used in games and toys. You can call it an "Executive Decision-maker." Remember, it is simply the electronic equivalent of flipping a coin. Anytime you might flip a coin, you can use this project instead.

Electronic dice

A two-possibility system such as in the last project is simple, but very limited. After all, there is always a 50% chance of guessing right on each trial. More sophisticated ESP testing calls for a wider range of possible outputs to choose from.

The last project was the electronic equivalent of flipping a coin. When you flip a coin, it's got to come up either heads or tails. But a common game die has six sides. The chance of randomly guessing correctly on any single trial is reduced to less than 17%. A long string of "hits" (correct predictions or identifications) by a test subject would certainly be much more impressive. The circuit shown in Fig. 4-2 is the electronic equivalent of a standard six-sided die. You can build 2 copies of this project to get a pair of electronic dice, for a total of 27 possible output combinations. A suitable parts list for this project appears in Table 4-2.

This project can be used for the same type of ESP tests as the preceding project. To test clairvoyance or foreknowledge, the subject should try to predict beforehand which output combination will turn up the next time the circuit is activated. To test telepathy, a second person uses the circuit in a separate room with no direct contact with the subject being tested. This second person activates the circuit, then concentrates on the displayed output while the subject attempts to read his mind. In either case, as with all ESP testing procedures, a large number of tests are needed to have any scientific validity. Anybody might guess right some of the time. In fact, it would be very surprising if a subject was always wrong over a moderately large number of tests.

In principle, this circuit is quite similar to the one shown in Fig. 4-1. A high-frequency oscillator steps through all of the possible output combinations at too high a speed to possibly be distinguishable. When the oscillator is stopped, the current output

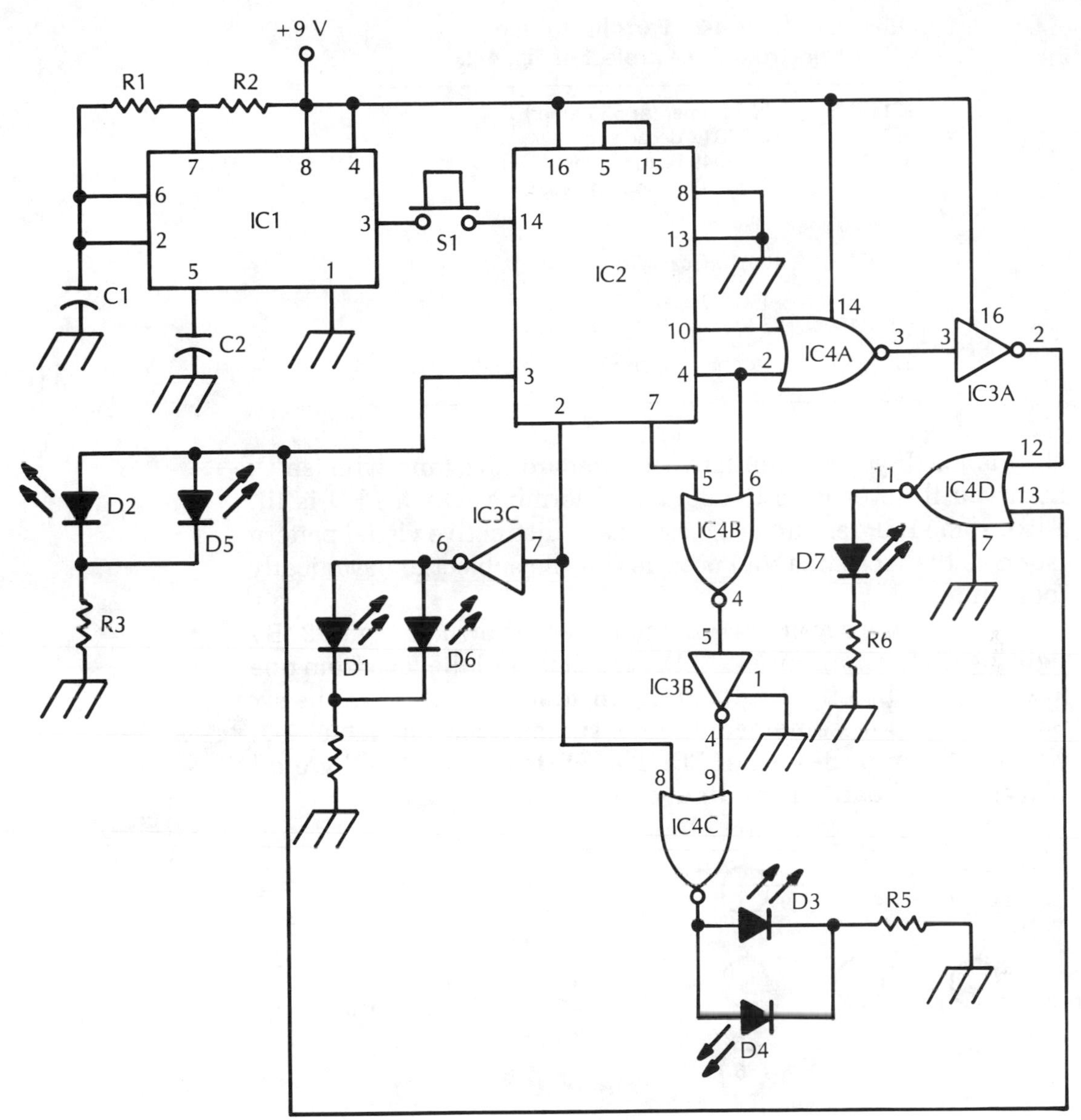

Fig. 4-2 Electronic dice project.

state is held and displayed. Push-button switch S1 permits the oscillations to get through when it is closed. Opening the switch (its normal state) blocks the oscillator signal preventing any change in the displayed output.

Seven LEDs are used to represent the six possible die faces. We can simplify the circuit somewhat and use a string of six LEDs, but this approach is more interesting and is easier to read

Table 4-2 Parts list for the electronic dice project of Fig. 4-2.

IC1	7555 timer (or 555 timer)
IC2	CD4017 decade counter
IC3	CD4049 hex inverter
IC4	CD4001 quad NOR gate
D1 – D7	LED
C1, C2	0.01-μF capacitor
R1, R2	1-kΩ, 0.25-W, 5% resistor
R3, R4	330-Ω, 0.25-W, 5% resistor
S1	Normally open SPST push-button switch

at a glance. If you're familiar with standard dice (and who isn't), then you'll never have to count to determine which LED is lit. Also, some ESP research suggests that a distinctive visual pattern (such as the faces of a die) is easier for the subject to psychically lock onto.

Arrange the seven LEDs in the pattern shown in Fig. 4-3. By lighting up the appropriate LEDs, any standard die face from one to six can be directly displayed, as illustrated in Fig. 4-4. The six possible output combinations are summarized in Table 4-3, where "*X*" indicates a lit LED. The LEDs marked "—" are off (dark) for that particular output combination.

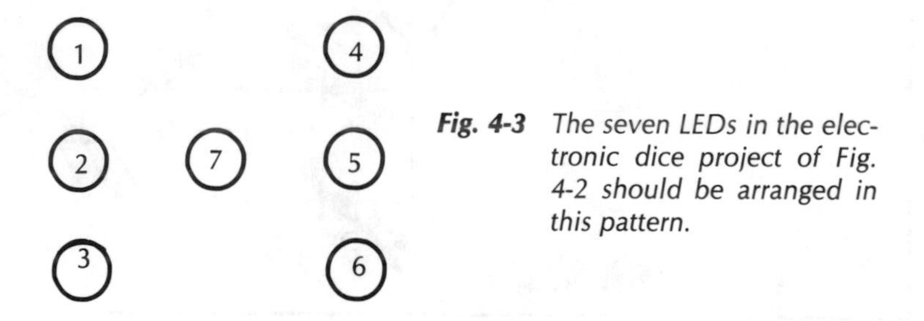

Fig. 4-3 The seven LEDs in the electronic dice project of Fig. 4-2 should be arranged in this pattern.

Notice that certain LED pairs are always used together. Either both members of the pair are lit or both are dark. There are three such pairs: D1 and D6, D2 and D5, and D3 and D4. Only D7 is used by itself. These pairs permit us to simplify the output gating quite a bit. Each pair of LEDs can be wired in parallel because they are always operating in unison.

Now, let's take a closer look at the circuitry of this project. IC1 is a 7555 CMOS timer wired as an astable multivibrator or

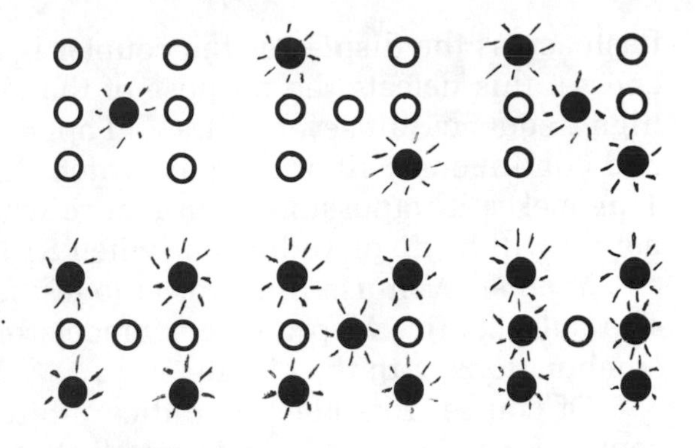

Fig. 4-4 *The seven LEDs in the electronic dice project of Fig. 4-2 can simulate any of the six standard die face patterns.*

Table 4-3 Summary of output patterns for the electronic dice project of Fig. 4-2.

Value	*D1*	*D2*	*D3*	*D4*	*D5*	*D6*	*D7*
1	—	—	—	—	—	—	X
2	X	—	—	—	—	X	—
3	X	—	—	—	—	X	X
4	X	—	X	X	—	X	—
5	X	—	X	X	—	X	X
6	X	X	X	X	X	X	—

rectangular-wave generator. Of course, you can substitute a standard 555 timer chip for the 7555 if you prefer. These two ICs are pin-for-pin compatible, and no changes are needed in this circuit if this substitution is made. The 7555 consumes less current than the standard 555.

The timer functions as a high-frequency clock. It is oscillating at all times, providing power is applied to the circuit. But its output pulses must pass through switch S1. When this switch is open (its normal condition), no pulses can get through to the next stage, and nothing happens. Closing switch S1 permits the pulses to pass through to a decade counter (IC2). While this chip is capable of counting from 1 to 10, it is wired here for a maximum count of 6 before resetting back to 1. The various stages of IC3 and IC4 gate the six active count output lines to light up the appropriate LEDs for each of the six possible output combinations, as shown in Fig. 4-4.

None of the component values are critical in this circuit. Feel free to make substitutions if they happen to be more convenient. The only restriction is that if any of IC1's frequency determining components (capacitor C1 and resistors R1 and R2) are given values that are too large, you will be able to see the various patterns flash across the display as the counter is being incremented. Of course, this defeats the purpose of the project. With a suitably high frequency, all seven LEDs will appear to be simultaneously and continuously lit as long as switch S1 is being held closed. This makes it impossible to influence what final output combination will be displayed when switch S1 is released.

You can even use this project to test for telekinetic abilities. The subject tries to produce a predetermined series of output combinations with the circuit.

Of course, this electronic dice project can also be used in games. It can be used in any game that calls for ordinary dice.

Random-number generator

A third version of the same basic idea is our next project, a random-number generator. Each time the circuit is activated a randomly selected digit from zero to nine is displayed. The subject being tested for ESP must determine the displayed output state for each test without using ordinary sensory means.

The circuitry for this project is illustrated in Fig. 4-5, and the suggested parts list is given in Table 4-4. Again, nothing is critical in this circuit. The exact component values are not important here. A simplified block diagram of this project is shown in Fig. 4-6. As you can see, there are three main sections to this circuit: the high-frequency clock (IC1 and its associated components), the decimal counter (IC2 and IC3), and the display (DIS1).

The clock section is a high-frequency astable multivibrator circuit built around the familiar 7555 (or 555) timer chip. The exact output frequency from this multivibrator is determined by the values of capacitor C1 and resistors R1 and R2. If these components are given very large values, the signal frequency will be slowed down too much to be practical for our purposes. Other than this minor restriction, almost any values can be used for these components. The component values listed in the parts list are simply those I happened to use in the prototype of this project. Use whatever you have on hand. Just to be safe, I'd keep the resistor below 47 kΩ and the capacitor below 0.1 μF. This should keep the signal frequency well above the visual limitations of our application.

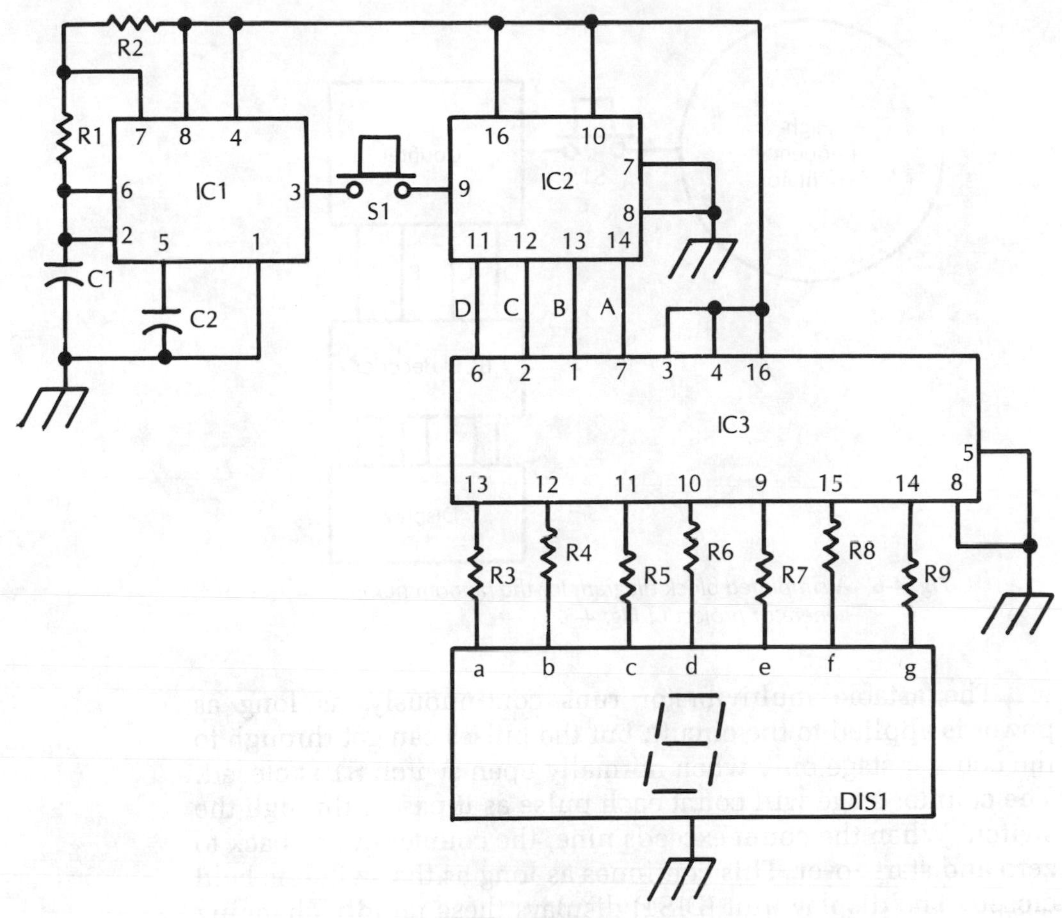

Fig. 4-5 Random-number generator project.

Table 4-4 Parts list for the random-number generator project of Fig. 4-5.

IC1	7555 timer (or 555 timer)
IC2	CD4518 BCD counter
IC3	CD4511 BCD to seven-segment decoder
DIS1	Seven-segment LED display (common cathode)
R1, R2	1-kΩ, 0.25-W, 5% resistor
R3 – R9	330-Ω, 0.25-W, 5% resistor
S1	Normally open SPST push-button switch

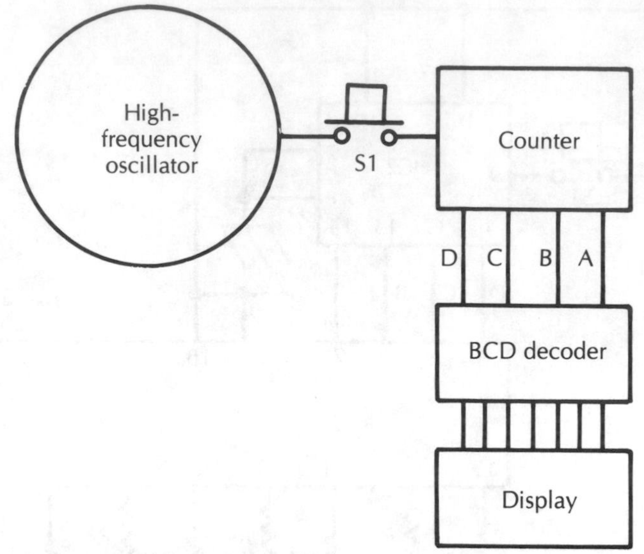

Fig. 4-6 *A simplified block diagram for the random number generator project of Fig. 4-5.*

The astable multivibrator runs continuously, as long as power is applied to the circuit, but the pulses can get through to the counter stage only when normally open switch S1 is closed. The counter stage will count each pulse as it passes through the switch. When the count exceeds nine, the counter cycles back to zero and starts over. This continues as long as the switch is held closed. The display unit (DIS1) displays these rapidly changing counts, but it changes so fast that the displayed numerals blur together, and it appears to display a continuous "8," possibly with a slight flicker.

IC2 is a BCD (binary-coded decimal) counter, and IC3 decodes the BCD values to properly display them in decimal form on the display unit; that is, IC3 "decides" which LED segments to light up for each count value.

When push-button switch S1 is released, the signal path is opened and no further pulses can get through to the counter stage. The last count value is displayed on the display unit—a value from zero to nine. The odds of randomly guessing the final display value of any given test are 10%.

The display unit is a standard seven-segment LED display unit. It must be of the common-cathode type (not common-anode). Other than that, there are no particular mandatory specifications for the display unit. Select the size, brightness, and color according to your personal tastes and based on just how you intend to use your project.

You can increase the brightness of the display somewhat by reducing the values of resistors R3 through R9. All seven of these current-limiting resistors should have the same value or the display will look odd and unbalanced. Alternately, you can decrease the brightness by increasing the value of these resistors. Do not use resistors with values lower than 100 Ω or the LED segments might be damaged by excessive current flow. If the resistor values are larger than 1 kΩ (1 000 Ω), the LED segments will probably be too dim to read.

If you build two of these circuits and mount the display units side by side, you can generate random two-digit numbers ranging from 00 to 99. The odds of randomly guessing the output of any given test would then be reduced to a mere 1%.

If you try the two-digit version, do not attempt to cut costs by using the same clock (IC1) for both digits. Both counters will see the same number of pulses, so the same digits will always be displayed on the display units, leaving you with just 10 possible combinations as in the original single-digit version of the project.

Unlike most other digital pulse counting circuits, you do not need to use a high-grade, bounce-free push-button switch for this application. If the switch bounces and a few extra false pulses are seen by the counter, it won't make any difference. If anything, it will increase the randomness of the output, which is the whole point of this project in the first place.

Of course, like the other projects presented so far in this chapter, this random-number generator project can also be used in games that require a random selection of numbers from zero to nine. Or, if you prefer, from 1 to 10—just consider a display of 0 to mean "10."

Manual telepathy tester

Perhaps in some tests of telepathic powers, it might not seem quite appropriate to focus on random values selected by an electronic circuit. You might prefer to have one person select entered values, while a second person (the subject being tested) tries to "read" the other person's mind to determine the selected value.

In this simple, but effective project, person 1 enters one of 16 possible switch combinations, then person 2 must enter the same combination. If person 2 succeeds, an LED lights up. If any one or more of the coding switches is in the wrong position, the LED will not light up.

The project must be built as two separate sections that can be physically and visually isolated from one another. Person 1 and person 2 should not be able to communicate with each other during the testing process, either visually or audibly.

Person 1's input board is shown in Fig. 4-7. Person 2's input board and the output encoding circuitry are illustrated in Fig. 4-8. Six interconnecting wires are required between the two sections of the circuit. A simple multiwire cable should do fine in

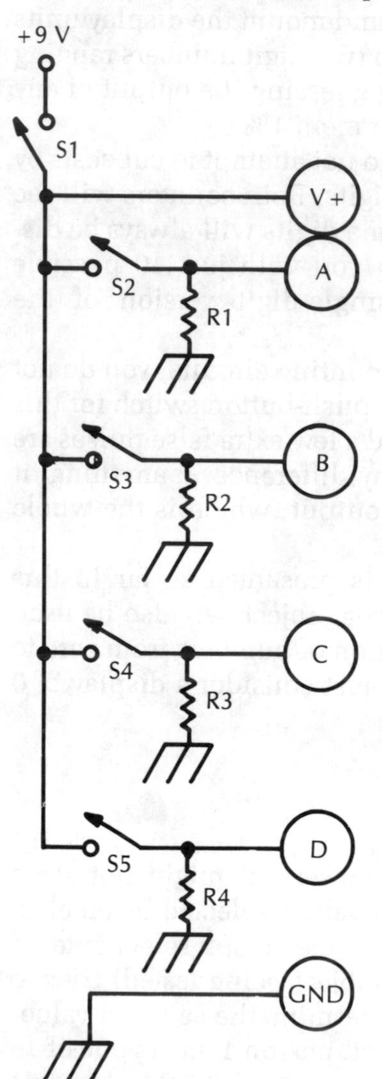

Fig. 4-7 *Input board for the manual telepathy tester project.*

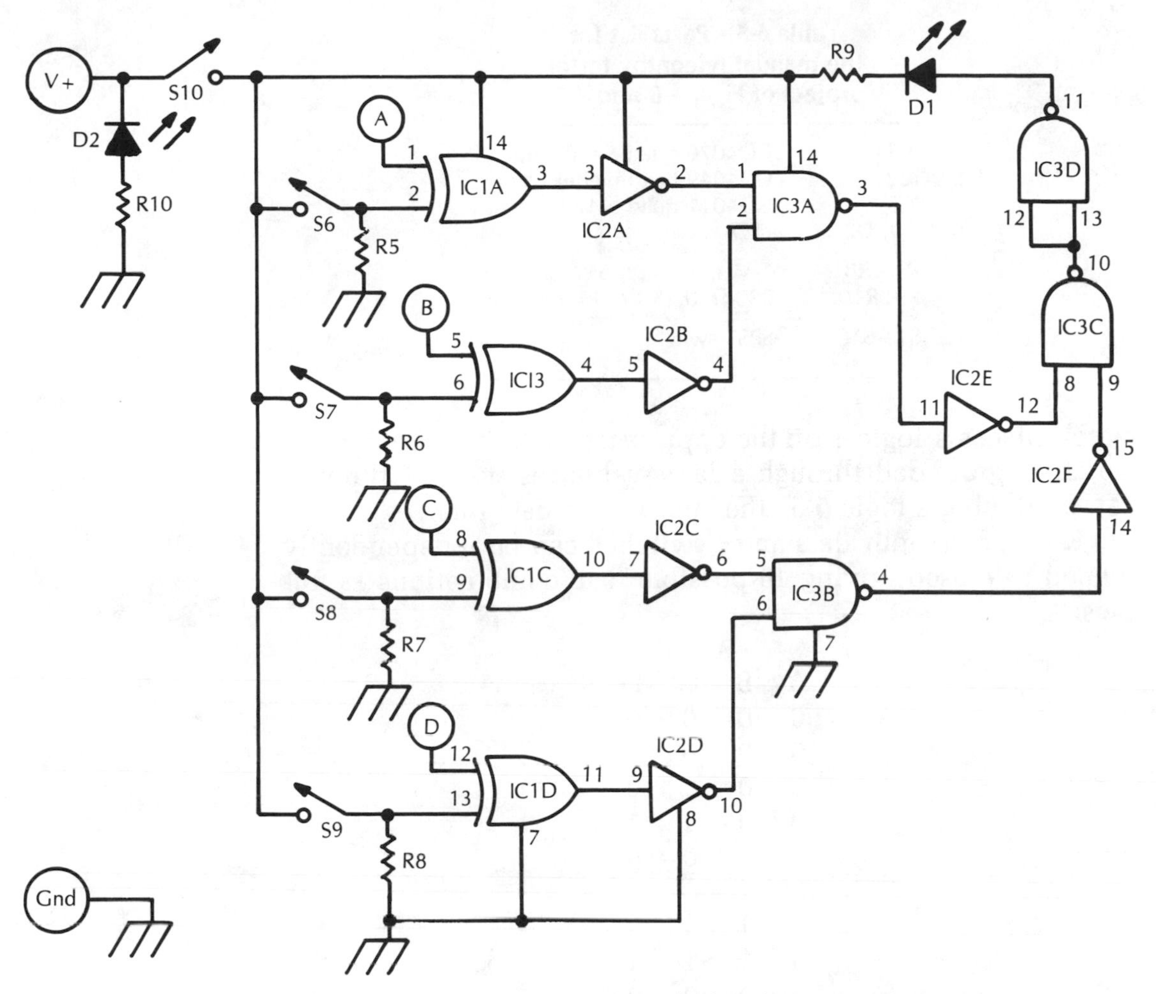

Fig. 4-8 *Output board for the manual telepathy tester project.*

this project. No special shielding is required unless the circuit is to be used in an environment with exceptionally strong interference signals. All of the signals carried by the interconnecting wires are steady-state dc voltages, which tend to be quite reliable and resistant to interference. The interconnecting wires carry the controlled power (V+), common ground (GnD), and four data lines (A, B, C, D). The complete parts list for this project is given in Table 4-5.

Initially, power switch S1 is left open, so no power is applied to the circuit. This helps prevent person 2 from cheating or possibly confusing the test results. Person 1 sets up the desired pattern by opening and closing switches S2 through S5. Each closed

Table 4-5 Parts list for the manual telepathy tester project of Figs. 4-6 and 4-7.

IC1	CD4070 quad X-OR gate
IC2	CD4049 hex inverter
IC3	CD4011 quad NAND gate
D1, D2	LED
R1 – R8	1-MΩ, 0.25-W, 5% resistor
R9, R10	390-Ω, 0.25-W, 5% resistor
S1 – S10	SPST switch

switch places a logic 1 on the appropriate data line. Each open switch is grounded through a large-value resistor (R1 through R4), producing a logic 0 on the appropriate data line.

Each of the four data entry switches can be independently opened or closed, giving 16 possible data combinations as follows:

A	B	C	D
0	0	0	0
0	0	0	1
0	0	1	0
0	0	1	1
0	1	0	0
0	1	0	1
0	1	1	0
0	1	1	1
1	0	0	0
1	0	0	1
1	0	1	0
1	0	1	1
1	1	0	0
1	1	0	1
1	1	1	0
1	1	1	1

Each 0 indicates an opened switch, and each 1 represents a closed switch.

Of course, this is simply a four-bit binary number, but the test subjects don't need to know or understand that. They just have to move each of the four switches to one position or the other. Most people will probably do best thinking of it as a visual pattern of switch positions. Therefore, slide switches are probably your best choice, because it is easy to see their positions at a glance. Toggle switches with large handles might also be suitable. However, I suggest avoiding push-on/push-off type push-button switches in this application.

Once person 1 has entered his selected pattern on switches S2 through S5, he closes power switch S1. This enables person 2's section of the circuit. Person 1 should not touch any of his switches until the test is completed.

Now, it is person 2's turn. For the time being, switch S10 is left open, so there is still no power applied to the gating circuit. Again, this prevents cheating and limits the possibility of confusing the test results. If S10 is closed prematurely, person 2 can just try every possible combination of switch positions until he finds the right one, which would completely sidestep the entire point of this project.

When LED2 lights up, person 2 concentrates on "receiving" the pattern in person 1's mind. Switches S6 through S9 are set to what person 2 thinks is the same pattern person 1 has set up. These switches work in exactly the same way as in person 1's input board. As before, each closed switch places a logic 1 on the appropriate data line. Each open switch is grounded through a large-value resistor (R5 through R8) producing a logic 0 on the appropriate data line. Thus, person 2 can choose from the same 16 possible combinations available to person 1.

When person 2 thinks he has the correct pattern set up, he closes switch S10, which applies power to the comparison gating circuitry. The two input patterns are compared by the gates of IC1, IC2, and IC3. If and only if all four switches are in the same positions for both sections of the circuit (person 1 and person 2 have entered the exact same pattern) then LED1 will light up. If one or more of person 2's switches are in the wrong position, the LED remains dark.

As with all ESP tests, there should be as large a number of tests as possible. The odds of randomly guessing the correct pattern are 1 in 16, or about 6%. A subject who scores more hits than misses is unlikely to be just lucky (although it is possible). Does your subject have telepathic powers? I'll leave that question entirely up to you.

Automated ESP tester

Another ESP tester circuit is shown in Fig. 4-9. A suitable parts list for this project is given in Table 4-6. In this project, a specific value is randomly selected by the circuit, and the test subject

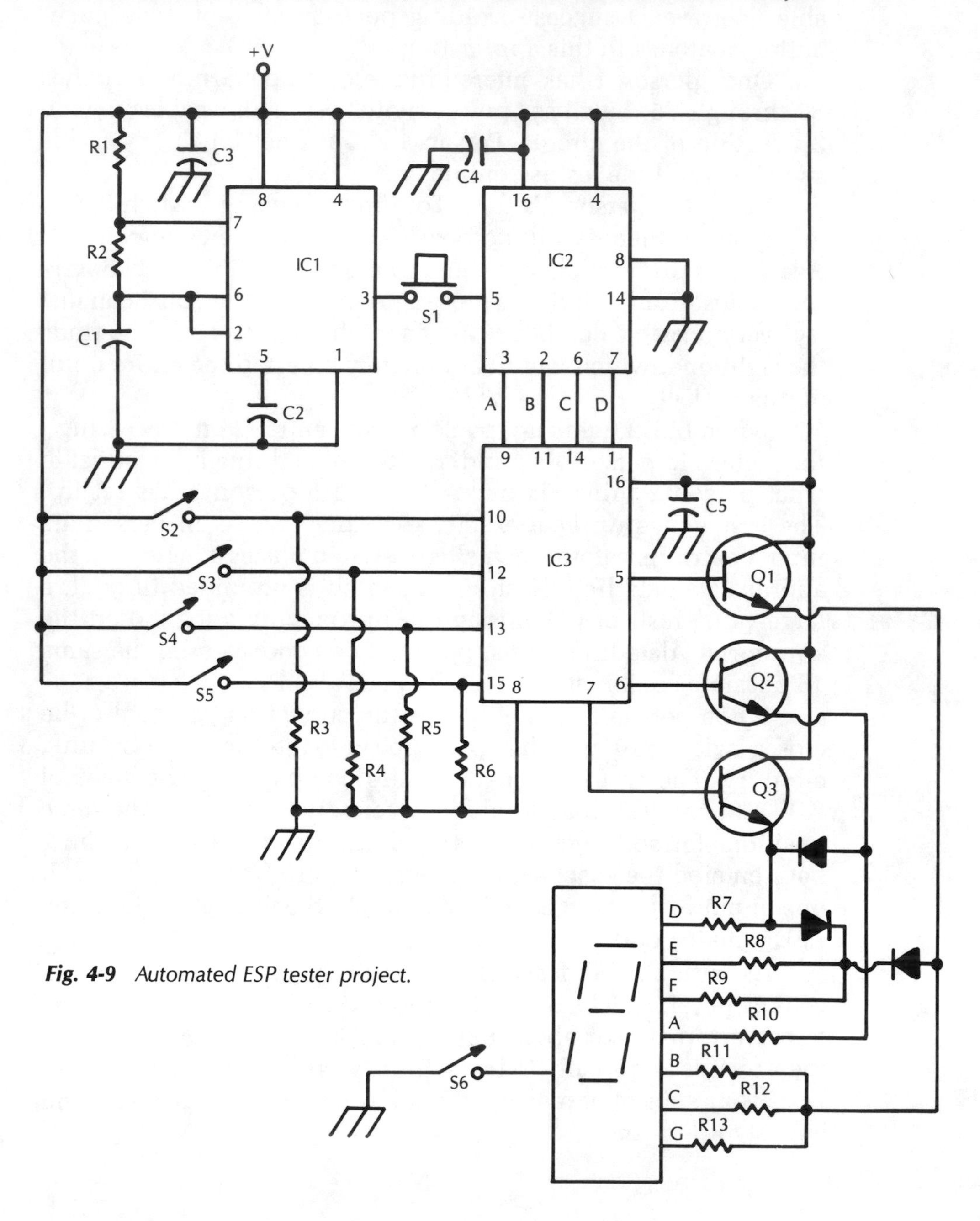

Fig. 4-9 *Automated ESP tester project.*

must enter the correct (unknown) value on four switches. The circuit then indicates if the subject was correct on a seven-segment LED display. This display unit is not used in the ordinary way to display digital numerals. Instead, it displays one of three letters: L for entered value is too low, H for entered value is too high, or C for entered value is correct. Of course, this project can also be used in a guessing game.

IC1 is a 7555 (or 555) timer IC in a standard astable multivibrator configuration and acts as a high-frequency clock. When push-button switch S1 is momentarily closed, the output pulses are counted by IC2. When switch S1 is released, the counting stops, and the last count value is retained by the circuit. The test subject enters his guessed value in binary form via switches S2 through S5. A closed switch counts as a 1 (or high), and an open switch is pulled low (0) through the appropriate grounding resistor (R3 through R6).

With four two-position (open or closed) switches, there are 16 possible combinations, which we can number from 0 to 15. The test subject does need to understand binary numbers. The subject just tries to determine the appropriate switch pattern; that is, which switches should be open and which should be closed.

If the test subject feels more comfortable thinking of actual numbers, it's just a matter of using the standard decimal/binary numbering conversion, as follows:

Decimal	Binary
0	0 0 0 0
1	0 0 0 1
2	0 0 1 0
3	0 0 1 1
4	0 1 0 0
5	0 1 0 1
6	0 1 1 0
7	0 1 1 1
8	1 0 0 0
9	1 0 0 1
10	1 0 1 0
11	1 0 1 1
12	1 1 0 0
13	1 1 0 1
14	1 1 1 0
15	1 1 1 1

Table 4-6 Parts list for the automated ESP tester project of Fig. 4-9.

IC1	7555 timer (or 555)
IC2	74C193 up/down counter
IC3	74C185 four-bit comparator
DIS1	Seven-segment LED display module (common cathode)
Q1 – Q3	npn transistor (2N2222, 2N3904, or similar)
D1 – D3	1N4148 diode
C1	0.047-μF capacitor
C2	0.01-μF capacitor
C3, C4, C5	0.1-μF capacitor
R1	39-kΩ, 0.25-W, 5% resistor
R2	22-kΩ, 0.25-W, 5% resistor
R3 – R6	1-MΩ, 0.25-W, 5% resistor
R7, R10 – R13	330-Ω, 0.25-W, 5% resistor
R8, R9	270-Ω, 0.25-W, 5% resistor
S1, S6	SPST NO push-button switch
S2 – S5	SPST switch

Notice that there are no other possible combinations.

Once the test subject has entered what he thinks is the correct pattern of switch positions, momentarily close switch S6. A letter is displayed. The test subject's entered value is compared to the stored count value by IC3. If an "L" is displayed, the value entered by the test subject is lower than the preselected count value. If an "H" is displayed, the value entered by the test subject is higher than the preselected count value. If the entered value is exactly the same as the predetermined count value, a "C" will be displayed, indicating that the answer was correct.

If you get the answer wrong, you can try entering a new pattern with switches S2 through S5 as before. Again, press switch S6 to find out if your new guess was any better than the last one. The preselected count value is held as long as switch S1 is not depressed again (or power to the circuit is not interrupted). To reset the unit to select a new value, momentarily close switch S1.

Of course, it is significant if a test subject is right more often than the laws of probability predict. It is equally significant if the test subject is wrong more often than the laws of probability predict. What do you suppose it means if a test subject is consistently too low or too high?

Used as a game, without ESP, you can try to guess the circuit's predetermined count value with as few guesses as possible. The "L" (too low) and "H" (too high) indications give you important clues, permitting you to use strategy. With a little practice, you'll be surprised at how few guesses you will need. This has nothing to do with ESP of course, but is pure logic. Whether you use this project to test for ESP or to improve logical thinking and strategy, it's a lot of fun.

❖ 5 Air ionizer

NONTRADITIONAL APPROACHES TO HEALTH ARE AN IMPORTANT part of New Age. New Age health practitioners do not accept the standard medical models of health and disease. In this chapter, we are concerned with air ionization and its effects on physical and emotional health.

Ions

What does ''air ionization'' mean? All matter is made up of tiny components known as atoms. Atoms, in turn, are made of still smaller units—electrons, protons, and neutrons. Each electron has a small negative electrical charge, and each proton has a similar small positive electrical charge. Neutrons are electrically neutral. An atom's protons and neutrons are bunched together into a cluster known as the ''nucleus.'' The electrons orbit around the nucleus, somewhat like planets around the sun.

Ordinarily, the electrical charge of an atom is neutral; that is, the negative charges (from electrons) within the atom exactly equal and cancel out the positive charges (from protons). This means that each ordinary atom has exactly as many electrons as it has protons. But under many conditions, an atom may lose one (or possibly more) of its orbiting electrons. This means that the atom now has more protons than electrons, so the total internal positive charges outweigh the total internal negative charges. The atom as a whole has a small positive electrical charge. Such an atom is called a ''positive ion.''

Similarly, it is possible for an atom to pick up one (or possibly more than one) extra electron, perhaps by ''stealing'' it from another nearby atom. With the extra electron (or electrons), the atom now has more electrons than protons, so the internal negative charges are greater than the internal positive charges. The atom as a whole has a small negative electrical charge. In this case we have a "negative ion."

Positive ions and negative ions frequently occur randomly in the molecules of the air because the loose atmospheric atoms are constantly moving about and often bump into one another, disturbing the orbit of some of the electrons. Usually there is an equal number of negative ions and positive ions, because when atom A picks up an extra electron from atom B, atom A becomes a negative ion, and atom B becomes a positive ion. Oxygen is an element that is easily ionized. Unless otherwise specified, from here on in this chapter when I refer to ions I am referring specifically to oxygen ions.

Under some atmospheric conditions, there might be a preponderance of either negative ions or positive ions in the air. For example, just before an electrical storm, there are a lot of positive ions in the air. After a thunderstorm, the atmosphere has an excess of negative ions. Well, so what? What does all this have to do with health and well-being?

The effects of air ionization on human beings

Have you ever noticed that just before a big thunderstorm you might feel rather tense and edgy? But after the storm passes, you suddenly feel somewhat uplifted. It feels so good to breathe the air after a nice, cleansing rain. If you haven't noticed the effects in yourself, watch how animals react to a storm in the air. Often just before the storm hits, they become agitated and perhaps hyperactive. After the storm, they appear quite relaxed.

What you are seeing are the effects of air ionization. The excess positive ions in the atmosphere before the storm cause tension, while the heavy concentrations of negative ions in the air after a storm passes enhance relaxation and a feeling of well-being.

Lightning has a very strong effect on air ionization, because lightning is nothing but a powerful electrical discharge in the atmosphere, producing a lot of free electrons that can combine with nearby neutral oxygen atoms to produce negative ions.

Of course, whenever people are tense and anxious, problems are likely to arise. The risk of accidents grows. There is an increase in violent crime and fights. Some studies have suggested an increase in crime of up to 20% due to such atmospheric effects.

Acting on such theories, scientists at the Hebrew University in Israel have designed a device called the Ionotron to enrich the air with negative ions. In that part of the world, dry winds, heavily laden with positive ions frequently blow across the desert. Throughout the Middle East such dry winds blowing in from across the Sahara are known as "khamsin." Similarly, other countries have words for such ill winds. To the Italians they are the "sirocco," while the French call them the "mistral." Even in the United States such positive ion winds are known as "chinook" in the Rockies, and as the Santa Ana winds in California. As the old saying goes: "An ill wind blows no good." Perhaps there was some folk wisdom about air ionization that led to this old saying.

There appears to be substantial evidence that air ionization has definite physical and psychological effects on human beings. Persons subjected to air rich in positive ions have difficulty relaxing or concentrating. They might even develop physiological symptoms such as nausea or migraine headaches.

While there is still some controversy over whether air ionization really causes such effects, for the most part, these theories seem to be gaining fairly wide acceptance in today's scientific community. For one thing, there is strong laboratory evidence. For example, it has been determined that a neurohormone called "serotonin" is produced in the bloodstream when a person is subjected to an atmosphere with a lot of positive ions. Serotonin is known to be a stress-related hormone. Too much serotonin in the body can produce many negative symptoms, including depression, inability to concentrate or relax, fatigue, shortness of breath, dizziness, and migraine headaches. Notice that these are the very same effects attributed to positive air ionization.

The negative effects of positive oxygen ions are more strongly established than are the benefits of negative oxygen ions. However, it has been experimentally demonstrated that exposure to negative oxygen ions can cause serotonin to break up into apparently harmless by-products. Of course, this relieves the serotonin-related symptoms mentioned above.

Some Russian researchers have concluded that the presence of negative ions in the air is actually essential to animal life. A study at the University of California demonstrated a link between air ionization and plant growth. When the number of negative ions in the air is significantly reduced, plant growth is noticeably stunted. Somehow the negative ions help plants grow, although a definitive explanation for this has not yet been found.

While there appears to be something to air ionization, some specific claims seem a bit less likely than others. Claims have been made that enriching the negative ion content of the air can increase intelligence, relieve allergies, help control viruses, and retard the growth of bacteria. None of these things have been scientifically disproven, but they have hardly been proven either. I find the claim that negative ionization retards the growth of bacteria to be a bit unlikely in the face of the fact that the same ionization aids plant and animal growth. Bacteria are microscopic life forms, not all that dissimilar to more-complex plants and animals. Why should bacteria react to negative air ionization in the opposite way as their larger and more-sophisticated relatives?

I also think the claims that negative ions directly improve or otherwise affect intelligence to be a little far-fetched, although there might be some link with the effective use of the intelligence the subject already has. After all, positive air ionization interferes with concentration, which is likely to result in a lower score on an intelligence test.

Cleaning the air

One health benefit of negative air ionization is well-established: clean air is obviously healthier than dirty air, filled with smoke, pollen, and dust particles. Negative air ions can actually help clean such particles out of the air. Negative ions act a little like dust (and other airborne particle) magnets. A negative ion will attach itself to such a particle, neutralizing it, and weighing it down. The neutralized particle falls harmlessly from the air. It simply falls to the ground instead of hanging around in the air waiting for someone to breathe it in.

Wouldn't it be wonderful if someone could devise a way to use this effect on a large scale to clean some of the pollution out of the atmosphere. If nothing else, the air ionization project presented in this chapter can help make a dent in the pollution levels in your immediate environment, at least, over the short term.

Negative-ion generators

Because ions are, by definition, an electrical phenomenon, it's only natural for electronics to get involved. (Remember, an ion is simply an electrically charged atom.) Special high-voltage circuits have been designed to emit negative ions into the air. The name for such a device is a negative-ion generator or an air ionizer. Because of the cleansing action of the negative ions (as described in the preceding section of this chapter), such devices are sometimes called electric air cleaners or something similar. Of course, commercial units are often given specialized names at the whim of the manufacturer.

All such devices work in a similar way. A high voltage is built up in some kind of probe. When this voltage is discharged it emits a strong burst of ions that attach themselves to the atoms in the atmosphere surrounding the probe, thus creating many negative ions. In effect, a negative-ion generator acts like a small, self-contained, and controllable source of lightning bolts. In nature, bolts of lightning are huge electrical discharges that create large quantities of negative ions in the atmosphere. This is why people feel so good and uplifted after a strong electrical storm. The negative ionization caused by the lightning bolts helps cleanse the atmosphere.

The probe of a negative-ion generator is usually a thin needle or a similar metallic item. The probe is given such a strong negative electrical charge (typically thousands of volts) that the excess electrons must escape into the surrounding atmosphere, forming negative ions with the neutral atoms in the immediate vicinity. These ionized atoms soon spread out into the atmosphere.

If you place your hand near the probe, you might be able to feel the "ion wind." The "ion wind" that can be felt near the tip of the probe is due to the rapidly moving ion flux from the generator. DO NOT TOUCH THE PROBE WHILE THE GENERATOR IS IN OPERATION. Remember, the probe is at a very high electrical potential. A dangerous or possibly even fatal electric shock is virtually guaranteed if you touch the probe!

Many air ionizers have the probe out in the open and fully exposed. This, of course, permits the device to exert its strongest effect by providing the most direct exposure to the atmosphere. But, in some respects, this is an inherently risky design. In the interest of safety, a well-designed negative-ion generator should be housed, making it impossible, or at least very difficult, for anyone to touch the charged probe while power is applied to the circuit. For instance, the probe can be placed under a perforated plastic dome, as illustrated in Fig. 5-1. Plenty of air atoms and molecules can get in and out through the small holes, but curious fingers can't. Another approach is to recess the housing for the probe, as shown in Fig. 5-2. Do not enclose the probe entirely or there will not be sufficient air flow for the negative-ion generator to do any good.

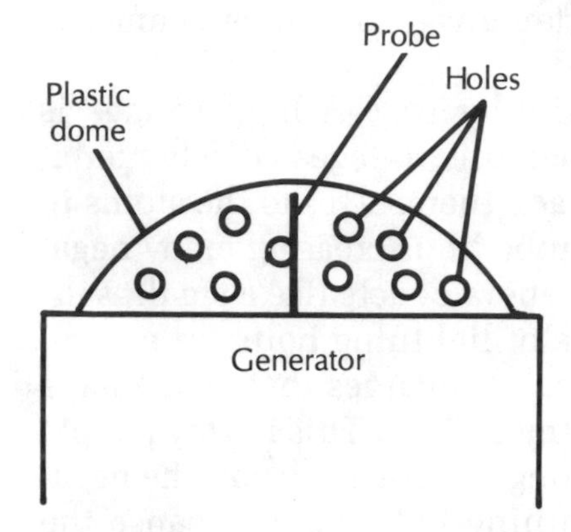

Fig. 5-1 For safety, the probe needle of a negative-ion generator can be placed under a perforated plastic dome.

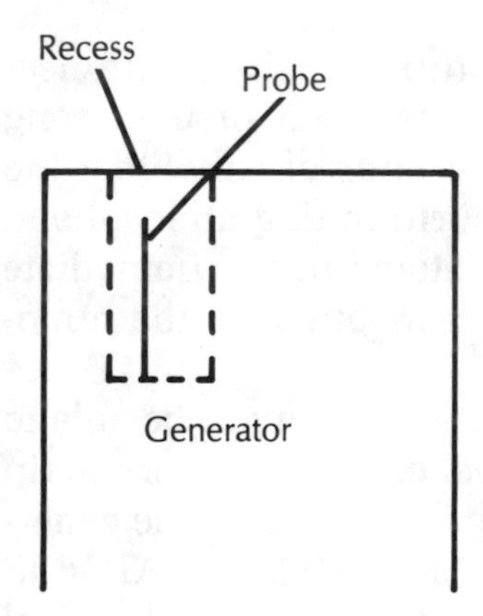

Fig. 5-2 The probe needle of a negative-ion generator can be recessed for safety.

An air ionizer project should not present a significant shock hazard if you just use a little common sense and make reasonable provisions for safety. You don't have to be a fanatic about it. Remember, you are guarding against somewhat freakish accidental possibilities. Don't be frightened away from this project, just don't ignore reasonable safety precautions.

The negative-ion generator should be placed several feet

away from the person who intends to enjoy the benefits of the enriched atmosphere provided by the generator. Of course this also helps limit the shock hazard, while offering the best possible effects. Perhaps you can mount the negative-ion generator high enough that no one can accidentally or easily touch the probe. Not surprisingly, this type of device will generally work better in a partially enclosed room. But good ventilation in the room is strongly recommended.

Negative-ion generator project

Figure 5-3 shows the circuitry for a negative-ion generator that you can build and experiment with. A suitable parts list for this project appears in Table 5-1. The circuit is really fairly simple. It is basically a high-voltage pulse generator. The system used here is known as the high-voltage corona discharge method, and it is probably the most commonly used approach in designing negative-ion generators.

As in so many of the projects in this book, IC1 is a 555 (or 7555) timer chip operated in the astable mode. Using the component values suggested in the parts list, the pulse frequency is about 65 Hz, with a duty cycle of a little less than 10%. In other words, we have a string of very narrow pulses.

Transistors Q1 and Q2 are a Darlington-pair amplifier for these pulses. Both must be npn transistors. The exact type number isn't critical, but Q2 must be a heavy-duty power transistor. These transistors are heavy conductors (especially Q3), so use adequate heat sinking. When in doubt, use more heat sinking. The only possible disadvantages to too much heat sinking are purely aesthetic. It's worth a little extra cost and circuit bulk to be safe.

The output from the amplifier stage is T1, which is actually a standard 12-V, 3-terminal automobile ignition coil. You probably won't find anything directly suitable at an electronics parts store, but you can readily find such units at automobile parts suppliers, or you might find a nice bargain at an automotive junk yard. Exact specifications for this device are not critical.

If you are not familiar with automotive ignition coils, you might have a little difficulty locating the high-V + output terminal. This is because in its normal use (in an automobile's engine) the connection is made with a heavy-gauge spark plug wire clip. The high-voltage terminal is inside a small tube-like projection at

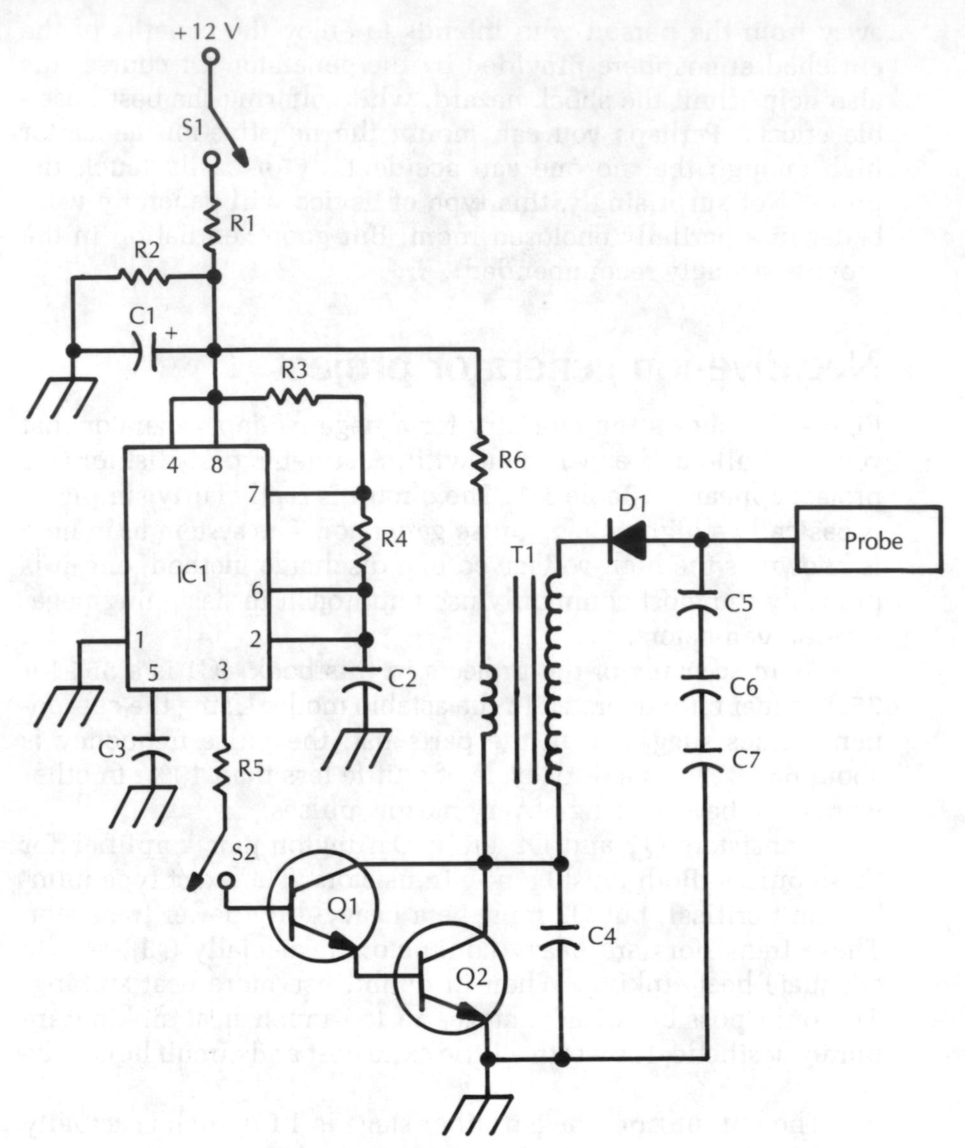

Fig. 5-3 *Negative-ion generator project.*

the top of the ignition coil unit. The stripped end of a heavy-gauge, well-insulated wire should be firmly inserted into the opening of this projection. Make sure there is a good, strong mechanical connection that won't pull free too easily. Except for the ends making the actual electrical connections, this wire should be very thickly insulated. Remember, it is carrying a dangerous level of electrical power.

Table 5-1 Parts list for the negative-ion generator project of Fig. 5-3.

IC1	7555 timer (or 555)
Q1	npn transistor (Radio Shack RS2030, 2N2102, or similar)
Q2	npn transistor (Radio Shack RS2041, GE-19, SK3027, MJE3055, or similar)
D1	45-kV diode array (ECG513 or similar)
T1	Automotive ignition coil (12-V, 3-terminal) (see text)
C1	500-μF, 35-V electrolytic capacitor
C2, C4	0.047-μF capacitor
C3	0.01-μF capacitor
C5, C7	470-pF, 6 000-V capacitor
C6	330-pF, 6 000-V capacitor
R1	1-kΩ, 10-W, 10% resistor
R2	10-Ω, 10-W, 10% resistor
R3	390-kΩ, 0.25-W, 5% resistor
R4	39-kΩ, 0.25-W, 5% resistor
R5	1-kΩ, 0.25-W, 5% resistor
R6	1.8-Ω, 10-W, 10% resistor
S1, S2	SPST switch

It is a good idea to completely enclose the entire ignition coil assembly in a plastic (or otherwise insulated) case of some sort for maximum protection against shock hazards. The ignition coil (T1) is much more of a risk for electrical shock than the ionization probe.

The coil assembly boosts the potential of the pulses considerably. The signal is half-wave rectified by D1, which is not an ordinary diode, but a 45-kV (45 000-V) high-voltage diode assembly. Make the connection to this diode assembly via standard high-voltage ignition wire. You will need a fairly hefty wire with good (and thick) insulation.

Of course, it is always very important to be careful about the polarity of all such components as diodes, but it is particularly critical for D1 in this project. If this diode assembly is installed backwards, your project will become a positive-ion generator. If you read the first half of this chapter you understand why that isn't very desirable.

Filter capacitors C5 through C7 must be high-voltage units. The total series capacitance works out to a little under 140 μF. Remember, the formula for capacitance in series is

$$\frac{1}{C_t} = \frac{1}{C_1} + \frac{1}{C_2} + \frac{1}{C_3}$$

The reason a single 150-μF capacitor is not used is because of the very high voltage involved at this point in the circuit. It is hard to find a single capacitor that can bear the full output voltage of the generator. By using three high-voltage capacitors, they can effectively share the voltage load. In this case, three capacitors are almost certainly less expensive than one to do the same job.

These three capacitors working together act like a standard filter capacitor in any half-wave power rectification circuit. Once filtered, the high-voltage output pulses from the generator circuit are fed to the probe. A common sewing needle is a good choice for the probe; it is an almost ideal size and shape.

While this circuit generates a very high voltage (about −6 kV to −9 kV (−6 000 V to −9 000 V)), the actual output current is quite low, so if someone accidentally touches the probe, it probably won't be fatal. But, whenever high power levels are involved there is always a risk of potentially fatal electrical shock, especially for anyone with any kind of heart ailment. At the very least, the probe does provide a substantial "kick" and can give a very painful shock, which can possibly cause serious injury. Be careful with this project.

It is strongly recommended that the probe (and all related wiring) be physically shielded, as discussed earlier in this chapter. But for the project to function usefully, there must be adequate air flow around the probe tip. A perforated dome shield or a recessed probe is a good idea.

Never use this project around children if they might be unsupervised, even for a moment. Children often find ways to defeat the most elaborate and carefully thought-out "foolproof" safety devices.

As an added precaution, two power switches are included in this project. Close switch S1 first to apply power to the pulse generator (IC1), then close switch S2 to permit the pulses to go through the amplifier and ignition coil (T1) to the probe.

Cautions and safety considerations

A negative-ion generator is not really a dangerous project, but it is, by necessity, a high-voltage device. Use all suitable precautions when working with such circuitry. When in doubt, increase the safety factors. It's far better to use too much insulation than too little. Make sure the circuit is well-shielded. Do not use an earth ground.

The creation of negative ions with a high-voltage pulse also increases the ozone (O_3) content of the air. Breathing significant quantities of ozone over extended periods is known to be potentially harmful to one's health. This negative-ion generator project is designed to minimize ozone production as much as possible, but you should be aware that there is some risk involved.

Just to be on the safe side, I recommend that you do not run this device continuously over long periods of time. Instead, turn it on for 15 to 30 minutes to build up the negative ionization, then turn it off for an hour or so. Besides preventing possible ozone buildup, this also reduces the possibility of overheating within the circuit, which can be a fire hazard when high voltages are involved.

It is also highly advisable to use this project only in moderately well-ventilated areas to further minimize the risk of possible ozone poisoning. Some of the generated negative ions will escape into the general atmosphere, but there will be so many of them within the immediate vicinity of the generator that you'll still be able to enjoy the benefits of negative air ionization.

Don't let these warnings scare you off from intelligently experimenting with this project or other negative-ion generators. The ozone risk and the shock hazard are actually fairly small, but they are real risks and deserve some intelligent consideration. It is always a good idea to consider the ''worst case'' scenario first, just to be on the safe side. If you follow the precautions suggested in this chapter, you shouldn't have much to worry about.

6 Biorhythms

IN THIS CHAPTER, I WILL DISCUSS A CONCEPT KNOWN AS *biorhythms*. Because of the nature of the material, this chapter is rather long on theory and a bit short on projects. Perhaps some readers can use the information presented here to design some new projects to test or use biorhythms.

It is interesting to notice that the traditional scientific community is vehemently opposed to some New Age concepts, is mildly (usually indirectly) opposed to others, and is utterly indifferent to some. This seems to be the case with biorhythms, at least in the United States. Despite an occasional pocket of interest, American scientists seem content to ignore the concept of biorhythms. In Europe there has been somewhat more interest and activity in this area, although biorhythm research is far from intensive anywhere. For one thing, the concept is inherently difficult (though not impossible) to test with proper scientific controls.

What are biorhythms?

The term "biorhythm" is sometimes used rather loosely. The roots of the word are easy enough to determine. The prefix "bio-" means "life" (as in "biology"), so a biorhythm is a life rhythm. There are many rhythmic cycles throughout the world of biology. We will look at some later in this chapter. For our purposes here, the term "biorhythm" refers to a very particular type of biological rhythm in human beings.

According to the theory of biorhythms, there is a sort of internal biological clock within each of us. This inner clock is very precise and regular. It operates with three simultaneous cycles or patterns, known as the physical cycle, the emotional cycle, and the intellectual cycle. Each of these cycles has its own unique rhythm. The period of each of these cycles is as follows: physical cycle, 23 days; emotional cycle, 28 days; intellectual cycle, 33 days.

Each of these cycles moves upwards and downwards symmetrically across a midpoint zero line, very much like the standard sine wave shown in Fig. 6-1. In fact, for those of us involved in electronics, it is very easy to visualize the three biorhythm cycles as three sine waves of differing frequencies, as illustrated in Fig. 6-2. Of course, the frequencies involved here are extremely low compared to those used in general electronics work, but the basic principle is the same.

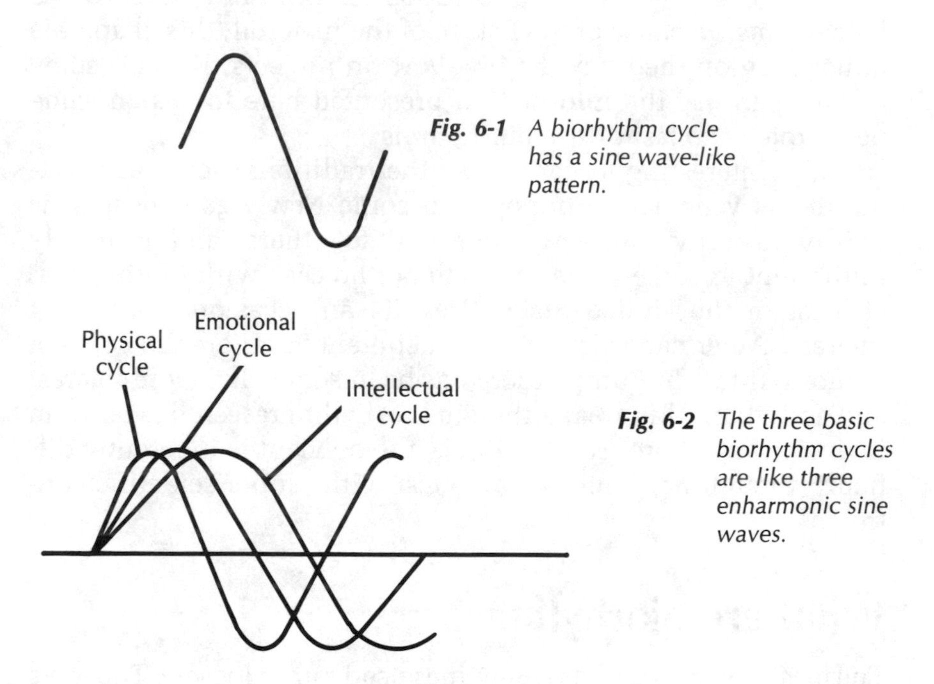

Fig. 6-1 *A biorhythm cycle has a sine wave-like pattern.*

Fig. 6-2 *The three basic biorhythm cycles are like three enharmonic sine waves.*

Each of these biorhythm cycles start at zero at birth, moving into a positive half-cycle. Because of their differing frequencies, the three cycles very quickly go out of synchronization. When one is moving upwards, the other two might be moving downwards. The three cycles won't resynchronize to the original starting condition (all zeros, going positive) for a period lasting 58 years, 66.5 days.

The names of the three basic biorhythm cycles are rather

self-explanatory. The physical cycle affects a number of physical factors and bodily functions, including such things as strength, coordination, speed, and resistance to disease, along with many others. Similarly, the emotional cycle affects mood, mental health, creativity, sensitivity, and the like. The intellectual cycle affects such factors as alertness, receptiveness to new ideas, logical thinking, and memory.

Remember, each of these cycles on any given day can be either negative or positive. To take just the physical cycle as an example, a person can be expected to be capable of his peak physical performance when his physical rhythm is near the top of its positive cycle. During the negative half-cycles, physical performance can be expected to be a little off.

But what is really important in terms of predicting performance is not the negative trough of the cycle, but the zero-line crossing points. Days when a cycle is crossing from positive to negative, or from negative to positive are known as *critical days*, and they are likely to reflect some sort of "confusion" related to the particular cycle. Accidents and other serious problems are most likely during critical days.

It is important to realize that because we have three asynchronous cycles operating here, when one cycle (say, the physical cycle) is critical, the other two cycles (the emotional cycle and the intellectual cycle) will probably be solidly positive or negative, and not critical. Less frequently, dual critical days can occur, when two of the cycles are simultaneously crossing the zero line, though not necessarily in the same direction. For example, on a typical critical day, the physical cycle might be critical going from positive to negative, while the intellectual cycle is critical going from negative to positive, while the emotional cycle is noncritical somewhere in its positive cycle. Of course, there are many other specific combinations, including dual critical days when both critical cycles are going in the same direction. A dual critical day, as you might suspect, is even more critical and potentially dangerous than a single critical day.

Even less common, but not rare, are triple critical days, when all three cycles are crossing through the zero-line at the same time, though probably not all in the same direction. If you are having a triple critical day, you really need to watch out.

Because the three biorhythm cycles are assumed to have very consistent and regular frequencies, and all three start at zero (going positive) on the day the subject is born, it is not difficult to calculate the exact point in each cycle for any specific day by adding up how many days the subject has been alive. This is tedious work to do manually, of course, but it is not at all difficult conceptually. Later in this chapter I will present a general BASIC computer program that can quickly do the mathematical "drudge work" for you in calculating biorhythms.

Rhythms in life

In the broadest sense, the theory of biorhythms is not that far-out or strange. Rhythmic patterns have been found throughout all of biology. Both plants and animals exhibit cyclic internal rhythms in abundance.

One of the most obvious biological cycles is the menstrual cycle in women, which lasts 28 days on the average. There is some variation in the length of the cycle from individual to individual, and many factors can lengthen or shorten the cycle at different times for a given individual. Some women are very regular in their menstrual cycle, while others are terribly irregular. Most are somewhere between these extremes. Most women are fairly regular, with occasional irregular deviations from their normal pattern. As a general rule, anything from about 26 days to 33 days is considered within the realm of normal. Menstrual cycles as long as 63 days are not unheard of.

It is not as widely known, because its symptoms are less obvious, but men also experience a monthly hormonal cycle, similar in many respects to a woman's menstrual cycle. Men, like women, can experience mood swings at certain times of the month. This is generally more likely to be shrugged off as just "being moody" because it is not linked to any overt, external symptom, such as menstruation.

Some men and women are extremely sensitive to the changes in their internal hormonal levels related to these cycles. Such people exhibit extreme mood swings and might even seem crazy to others. Very extreme cases in women have been studied as PMS (premenstrual syndrome). To the best of my knowledge, there hasn't been any research into the question of whether men might experience something akin to PMS due to their own hormonal cycles.

Some men and women aren't really emotionally affected to any noticeable degree by their hormonal cycles. Perhaps they just feel a little "off" on certain days, but that's all. Most men and women, of course, fall somewhere between these extremes.

The male hormonal cycle, like the female menstruation cycle, has an average period of 28 days, but it can vary considerably in response to many factors. While it hasn't been the subject of a lot of research, the existence of the male hormonal cycle has been pretty firmly established.

There are many other cycles in life. Many biological cycles are circadian. This imposing looking word is from the Latin and breaks down to "around (circa) the day (dies)." In other words, a *circadian cycle* is a daily cycle with a period of (about) 24 hours. Many circadian cycles are subtle enough that we usually don't notice them, but they have been well-demonstrated to be quite real. For example, there are definite daily rhythmic fluctuations in temperature, heart activity, blood composition, and many other bodily functions.

Our patterns of sleeping and waking hours represent a good example of a circadian cycle. The nominal pattern for humans is 16 waking hours, followed by 8 hours of sleep, then the cycle repeats. We can deliberately deviate from this cycle (sometimes such deviations aren't really the subject's choice), but at a cost. When our sleep/waking cycle is disrupted, we don't function as well, and we often become more prone to disease and stress.

But it is not just the 2:1 waking to sleep ratio that is important. The timing is also fairly critical. We seem to have internal clocks that "know" when it is time to sleep and when it is time to wake up. While it is a controversial idea, I believe that there are morning people and night people who can only reset their internal clocks with great difficulty. I know I am a night person, and I never could get used to getting up early in the morning, no matter what time I went to bed the night before. I find I work best and most efficiently if I go to bed a couple of hours later than "normal" and shift my entire schedule 2 to 3 hours from the "normal" average. (My parents confirm that I exhibited this shifted pattern even as an infant.) Other people are morning people, who like to get up at the crack of dawn and can't keep their eyes open past 10 P.M. Again, most people fall between the extremes and can learn to reset their internal clock (within limits) without too much difficulty.

However, it has been scientifically demonstrated that it is harmful to change the natural rhythm too much. Several studies have indicated that night-shift workers (with the ordinary sleeping/waking pattern reversed) are much more likely to make mistakes and have accidents. In other words, they just don't function as well on this "backwards" schedule. This is of considerable concern to industries that must operate 24 hours a day. No really good solution has yet been found.

By far the worst ''solution'' is a shifting schedule. Many jobs require that workers take a day shift for a few weeks then an evening shift for a few weeks, then a late night shift for a few weeks, then returning to the day shift for another cycle. This is probably one of the very best ways to minimize worker efficiency and health and to maximize stress. The body never gets a chance to stabilize into any consistent rhythm, much less a natural one.

Another approach that has been scientifically demonstrated to lower efficiency and increase stress is a split shift. The 8 hours of sleep per 24-hour period can't be broken up into any pattern, such as 8 hours of waking time, 4 hours of sleep, 8 more hours of waking time, and 4 more hours of sleep. Yes, this adds up to the same 16 hours of waking time, and 8 hours of sleeping time in the same 24-hour period, but the subject will experience considerable stress and lowered efficiency, and will almost certainly feel unrested.

Minor changes in the waking/sleeping pattern can be accommodated over time. It takes a week or two (and possibly longer) for the body to get used to any significant change. If the change is too great, the body never does fully adapt to the altered cycle.

Some people are fairly flexible about this cycle, while others have very rigid needs. The 16:8 ratio is an average. Some people need a little less sleep per 24-hour period, while others need a little more. External conditions can also change the cyclic pattern somewhat at certain times.

When test subjects are isolated from all normal cues of day or night, such as in a cave, they tend to settle into a 24- to 25-hour daily pattern without any external clocks. Some people tend to lock onto a slightly shorter than normal cycle, but it is more common to lock onto a slightly longer than normal cycle. However, it is significant that almost everyone in such experiments does settle down to a daily cycle that is fairly close to 24 hours.

This raises some interesting questions for extended space travel and the possible colonization of other planets that may have day/night periods considerably shorter than or longer than the 24-hour days of earth.

Biorhythm enthusiasts often speak of jet lag, which is a well-established syndrome related to the disruption of biological cycles or rhythms. Jet lag is a phenomenon that is well-known to frequent air travelers. When flying from one time zone to another, the body's rhythms are suddenly put out of synchronization with all environmental cues. Until he's had time to adjust and reset his internal clock, the jet lag sufferer doesn't feel hungry at mealtimes or sleepy at bedtime, and is certainly not ready to go to work (or perform other activities) when it is time. He is very likely to be moody and significantly less efficient and capable than normal, especially in areas of mental performance.

The effects of jet lag are sufficiently acknowledged today so that presidents and other high officials try, whenever possible, to include a day or two of rest time before undertaking any serious meetings or work when traveling to countries in distant time zones. This is not just a paid vacation, but an attempt to overcome the effects of jet lag.

For obvious reasons the problems of jet lag are particularly important to pilots and other key members of airline flight crews. A number of studies have been made in this area. To minimize jet lag effects as much as possible, there is often an attempt to keep the pilot on a home-time (rather than local-time) schedule as much as possible. No one wants an airline pilot to be excessively moody or below normal in his mental abilities on the return flight.

In biological terms, it is clear that time is a real and important factor in normal functioning. Biology is filled with rhythmic cycles, so the basic idea of biorhythms certainly seems reasonable enough.

Flaws in the reasoning

Ironically, many of the same things that are used as arguments for the biorhythm theory also offer arguments against that same theory. To be useful and meaningful as a predictive tool, biorhythms are assumed to be universally consistent and regular. That is, the physical cycle, for example, is always 23 days long for everybody throughout their entire lives. It never slows down to a 24-day cycle or speeds up to a 22-day cycle. It is never influenced in any way by any external factors.

But this is totally unlike all other known rhythmic cycles in biology. In fact, jet lag, which is almost always mentioned in arguments in support of biorhythm theory, stems directly from the adjustment of internal rhythms to external factors when moving from time zone to time zone. The menstrual cycle, which is also almost always mentioned in discussions of biorhythms, is actually highly variable and sensitive to influences from many possible sources, ranging from environmental changes to emotional or physical stress. So, while rhythmic cycles are common throughout biology, the consistent and invariable cycle lengths described by biorhythm theory certainly appear to be very unique.

While I have not been able to obtain copies of the original raw data, it appears that the lengths of the three biorhythm cycles were determined by studies involving statistical averages; they are not absolute lengths as assumed by modern biorhythm theory. Any given cycle may be a little shorter or a little longer. Without extensive raw data, it is impossible to determine how much variation might be considered normal.

My personal opinion is that there probably is something to biorhythm theory, at least in general. Recurring internal cycles affecting physical, emotional, and intellectual conditions and functioning seem quite plausible. But the cycle lengths are almost certainly far more variable than the current theory allows. Of course, this makes the system far less useful as a predictive tool. In fact, the theory seems to become little more than a curiosity with very limited (if any) practical application.

Another questionable aspect of biorhythm theory is the assumption that all three biorhythm cycles begin at zero, moving positive at the time of birth. Why? Biologically, there really isn't that much difference between the day before birth and the day after birth, except that after birth the baby is breathing on its own rather than obtaining its oxygen from the mother's blood through the placenta. Supporters of biorhythm theory argue that the birth trauma is the explanation. The trauma of being born is so strong that all internal clocks are reset to zero.

The trouble with this explanation is that it virtually destroys the rest of biorhythm theory. The birth trauma explanation states that external factors (the severe trauma of being born) can significantly influence the internal cycles that biorhythm theory requires to be universally consistent and invariable. Why don't severe traumas later in life affect the biorhythms? It doesn't explain anything to insist that the birth trauma is so much stronger than virtually all later life traumas. OK, so a later trauma may not reset the internal clocks to all zeroes like the birth trauma supposedly does, but surely they should have some effect on the cyclic patterns. And what of people who have had near-death experiences? Surely this would be a trauma on a level comparable to the birth trauma. How do such experiences affect biorhythms?

Assuming, as biorhythm theory does, that the cycles are all consistent and invariable, then everyone who lives 58 years and 66 to 67 days should experience some significant effect on that day as the supercritical condition of birth (all cycles at zero, going positive) is repeated. If there is anything to biorhythms as a useful theory, there should be some noticeable effect of some kind. Yet, to the best of my knowledge, no culture places any particular importance or mythic value on the fifty-eighth year of life. No universal symptoms have been noticed in that year, despite biorhythms. But other years and dates have been held as critical by various cultures. For example, in our culture, the thirtieth and fortieth birthdays are assumed to be crisis points for many people, even though these dates aren't particularly significant to biorhythms.

Please notice that raising these questions does not in any way disprove the theory of biorhythms. But until such questions are satisfactorily answered, it is probably premature to place too much faith in such a theory. Like so many other New Age concepts, the theory of biorhythms is certainly intriguing and worth scientific consideration, but as it stands it is flawed, with many gaps in the available information. (Such gaps are all the more reason for more serious scientific research.)

Some historical background

So where did the idea of biorhythms come from? Unlike many (but not all) other New Age concepts, biorhythms did not originate from any form of mysticism. Rather, the biorhythms idea was created like any scientific theory—from organized observation. The biorhythm theory was arrived at independently and more or less simultaneously around 1900 by two scientists, Dr. Hermann Swoboda of Vienna and Dr. Wilhcm Flicss of Berlin. There is absolutely no evidence that these two men had any idea of the other's work, at least in the early stages of their research. It is certainly significant that two scientists independently came up with the concept. If there is nothing to biorhythms, if it is pure fancy, the odds against such parallel development are certainly fairly large.

These two "fathers of biorhythms" were not quacks from the far fringes of science. They were both highly respected in their fields. Dr. Swoboda was a professor of psychology at the University of Vienna. Dr. Fliess was a well-respected nose and throat specialist. He later went on to become president of the German Academy of Sciences.

Of course, their reputations don't, in themselves, prove their theories. Many eminent scientists have been led down a false path of research, where exciting early results later fizzled out, forcing the abandonment of the hypothesis. Has this happened in the case of biorhythms? That is strictly a matter of opinion. Conclusive proof for or against the theory of biorhythms still hasn't been found.

Swoboda, as a psychologist, started to notice a rhythmic pattern in his patients' dreams, ideas, and creative impulses. New mothers tended to show anxiety about their infants just before or during what later biorhythm theory labeled "critical days." Swoboda kept careful records of such patterns. He also considered physiological factors such as fevers, pain, and swelling of tissues. His attention was drawn towards swellings with no apparent explanation. He even kept track of recurring illnesses, bouts of asthma, and even heart attacks.

Statistically analyzing his data he discovered evidence of definite 23- and 28-day cycles—that is, the physical and emotional biorhythm cycles. Apparently, these cycles were very regular and predictable.

Because I don't have access to Swoboda's raw data, I have to wonder whether the cycle lengths are absolute, or are they statistical averages? That is, did every single subject exhibit a fixed physical cycle of exactly 23 days? Or did some people have slightly shorter or longer cycles, with the average working out to 23 days? Modern biorhythm theory, of course, assumes that unlike all other known biological rhythmic cycles, the biorhythm cycle lengths are 100% consistent throughout life and completely identical for all subjects with no variation. As far as I have been able to determine, no hard data to support this claim has been published.

In any event, Swoboda's findings were certainly striking, and did not appear to be a fluke. He began assembling his data into what became known as biorhythm theory and wrote several scientific books discussing the subject.

About the same time as Swoboda was making his discoveries, Fliess was also studying carefully kept records of his patients' symptoms. He too discovered compelling evidence of

23-day and 28-day cycles. Notice that both researchers not only discovered biorhythms, but they also independently agreed on the cycle lengths, although there is still the question of whether they were considering statistical averages. Even if their numbers were derived from statistical averages, this is still impressive evidence that there really might be something to biorhythms. Swoboda seemed to find a lot of local support for his theories, but Fliess did not. This is interesting, because the two men were espousing almost the same theory.

The 33-day intellectual biorhythm cycle was discovered by a third researcher, an Austrian doctor of engineering named Alfred Teltscher. He taught at Innsbruck, and he decided to study the records of his students' performance to determine if there was a cycle of intellectual performance as Swoboda and Fliess had suggested there was for physical and emotional performance. Of course, I would not be mentioning Teltscher unless he obtained positive results. Sure enough, he found a regular 33-day cycle in the students' performance in class. This cycle was later confirmed by other studies.

We cannot fairly analyze the history of biorhythm theory without some consideration of the times. The late nineteenth and early twentieth centuries, when the theory was developed, was the height of "scientism," the almost fanatic belief that the scientific method and the scientific approach can solve any and all problems of life and the universe. The mechanical "clockwork" model of the universe held full sway then. The universe (and thus, everything within it) was thought to be a sort of machine, behaving in predictable and consistent ways. Later developments in relativity and especially quantum physics shot gaping holes in this nice, neat image, though many continue to cling to it, even today, insisting they are being "scientific." Naturally, the whole idea of biorhythms made a lot of sense in this context. Could such a theory originate today with all the upsets from quantum physics?

Of course, none of this proves or disproves anything about the ultimate reality of biorhythms. But it does raise some questions that need to be answered.

For whatever reasons, modern biorhythm theory has found a lot more acceptance in Europe than in the United States. Many official European organizations use or study biorhythms, while the entire subject is more or less ignored in the United States. But acceptance of biorhythm theory is far from universal even in

Europe, especially among serious scientists. The subject remains controversial, primarily because of the lack of hard evidence.

Given the nature of the subject, hard evidence of biorhythms is not easy to obtain. There have been relatively few extensive studies in this area. The few extensive studies and numerous small-scale studies into biorhythms have certainly had intriguing results, but a lot more major research is required before the questions about biorhythms can ever be scientifically settled.

Biorhythm calculation program

While it is extremely difficult to set up a tight, scientific study of biorhythms with all appropriate safeguards against inaccuracies and questionable procedures, it is not too hard to check out biorhythms on a small scale. Just calculate your biorhythm values and compare them to what is actually going on in your life (physical, emotional, and intellectual performance) during the study.

How do you find the appropriate biorhythm values for a given date? Simply count how many days you (or some other subject) have been alive from the date of birth to the date(s) of interest. That is easy enough in concept, but in actual practice it is pretty tedious, and it's all too easy to make mistakes. Also, matters are complicated by keeping track of leap years. Not many people want to be bothered with all the pencil work.

Fortunately, these days there is an easy answer—use a computer. Such repetitious mathematical calculations are just the sort of thing a computer is best at. Our next project, therefore, is not an electronic circuit, but a computer program. This biorhythm calculation program is written in BASIC and is as generalized as possible. This program should run on almost any standard personal computer that uses BASIC. Some machines use dialects of BASIC that might require a few minor modifications, but for the most part, the program should run just fine as written. No special or unusual commands are used.

This is not the most concise or elegant program possible. Any readers with programming experience should be able to improve the program as presented here. This is the result of a deliberate choice. Rather than using fancy and tidy programming tricks, I wanted to set up this program to make it as easy as possible to follow. Many remark statements (preceded by an apostrophe (')) are included for the same run. These remark lines can be deleted without altering the program.

We will look at each step in the program, so you can fully understand it. The entire program is summarized in the flowchart of Fig. 6-3. The complete program itself is given in Table 6-1. All lines beginning with an apostrophe are remarks. These are just handy programmer notes that are ignored by the computer when the program is running. In our discussion we will also ignore the remarks.

Line 50 dimensions (DIM) an array. An array is a multiposition variable. In this case, we are setting up a 12-stage single-dimension array, ranging from MD(1) to MD(12). Each "position" can hold a unique and independent value. We will use this array to hold the maximum number of days for each month of the year. I selected "MD" to stand for "month days." Lines 60 through 100 display an introductory message on your monitor screen.

Lines 120 through 140 load the array (MD(x)) with the appropriate values. These values are read from the DATA statements that appear late in the program, specifically lines 1000 and 1010. The array "memorizes" these 12 values for later use throughout the program.

The next section of the program prompts the user to enter his or her birth date, which, of course, is the starting point for calculating the biorhythm values. Each part of the birth date must be entered separately. Lines 170 and 180 ask for the birth year (YR). The program, as it is written here, is set up to accept birth years between 1900 and 1999 (lines 190 through 230). Of course, the program can be adapted to accept a wider range of years. If the entered year falls outside the acceptable range, the message, "Invalid year" is displayed, and the prompt is repeated until a valid year value is entered.

Notice that line 190 includes the statement, "YR = INT(YR)." This statement (with suitable variable names) is used throughout the program to prevent anyone from accidentally (or deliberately) entering a fractional value, such as 92.38. Everything to the right of the decimal point is cut off and ignored by the INT statement. "INT," as you might have guessed, is short for "integer."

The year can be entered as the complete four digits (such as 1992) or by the last two digits (92). In the second case, the program adds 1900 to the two-digit value to form the full year and displays the assumed year.

Lines 240 through 260 ask for the birth month and check it for validity. Of course, the month value (MO) must be a whole number between 1 and 12, as there are 12 months in a year.

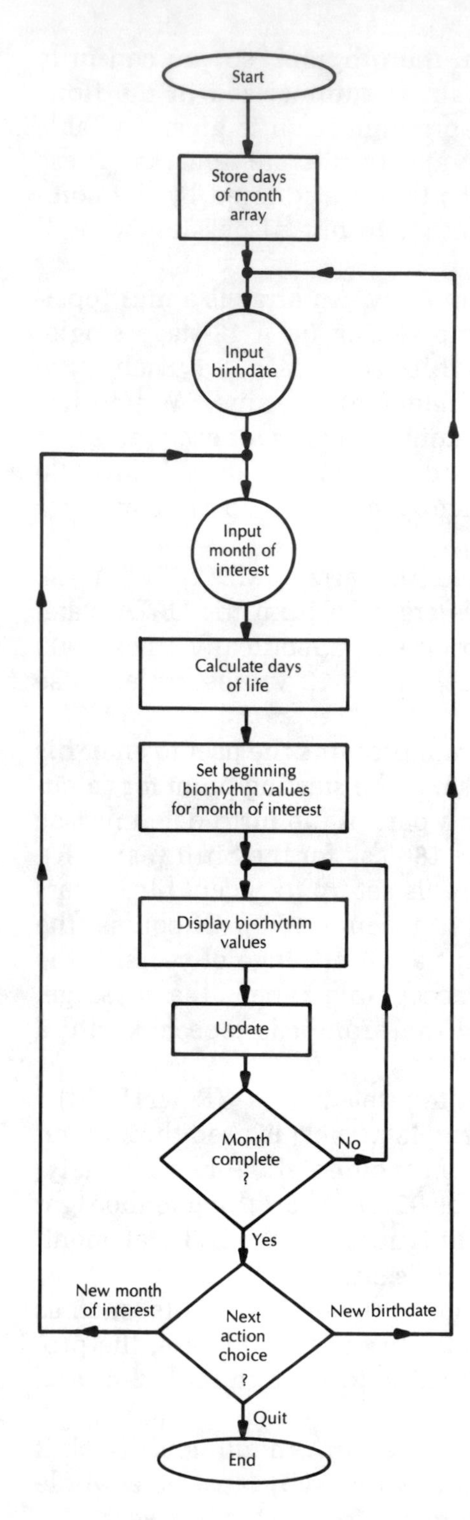

Fig. 6-3 *A flow chart of the biorhythm calculation program.*

Table 6-1 Biorhythm calculation program

```
10  ' BIORHYTHM CALCULATOR
20  ' BY DELTON T. HORN
30  ' COPYRIGHT 1991
40  '
50  DIM MD(12)
60  CLS
70  PRINT:PRINT:PRINT
80  PRINT "CALCULATE YOUR"
90  PRINT "BIORHYTHMS!"
100  PRINT:PRINT
110  ' STORE DAYS OF MONTH
120  FOR X=1 TO 12
130  READ A:MD(X) = A
140  NEXT X
150  ' INPUT BIRTH DATE
160  ' MO = MONTH, DA = DAY, YR = YEAR
170  PRINT "INPUT BIRTH YEAR";
180  INPUT YR
190  YR=INT(YR):IF YR>1900 GOTO 230
200  IF YR<100 GOTO 220
210  PRINT "INVALID YEAR":GOTO 170
220  YR=YR+1900:PRINT YR;"ASSUMED"
230  IF YR>2000 GOTO 210
240  PRINT "ENTER MONTH OF BIRTH (1-12)"
250  INPUT MO
260  MO=INT(MO):IF MO<1 OR MO>12 GOTO 240
270  PRINT "ENTER BIRTH DAY (1-";MD(MO);")"
280  INPUT DA
290  DA=INT(DA):IF DA<1 OR DA >MD(MO) GOTO 270
300  IF MO=2 AND DA=29 GOTO 1070
310  '
320  ' INPUT MONTH FOR BIORHYTHMS
330  ' YB = BIORHYTHM YEAR
340  ' MB = BIORHYTHM MONTH
350  PRINT "INPUT YEAR OF INTEREST"
360  INPUT YB
370  YB=INT(YB):IF YB>99 GOTO 320
380  YB=YB+1900:PRINT YB;"ASSUMED"
390  IF YB<YR OR YB>2100 THEN PRINT "INVALID YEAR"GOTO 350
400  PRINT "INPUT MONTH OF INTEREST (1-12)"
410  INPUT MB
420  MB=INT(MB):IF MB<1 OR MB>12 THEN PRINT "INVALID MONTH":
       GOTO 400
430  ' CALCULATE BIORHYTHMS
440  PRINT:PRINT:PRINT
450  PRINT "* * * CALCULATING * * *"
460  ' ADJUST MONTHS
470  YX=YB:IF MB>MO THEN MX=MB-MO: GOTO 520
480  YX=YX-1:MX=12-(MO-MB)
490  ' CALCULATE YEARS
500  '
510  ' COUNT OF DAYS ALIVE
```

Table 6-1 Continued.

```
520  DC=INT (YX*365.25)
530  ' CALCULATE DAYS IN BIRTH MONTH
540  DC=DC+(MD(MO)-DA)
550  ' CALCULATE DIFFERENCES IN MONTHS
560  IF MX<1 THEN MY=MO ELSE MY=MB:MX=ABS(MX)
570  MD(2)=28:LP=YB/4:IF LP=INT(LP) THEN MD(2)=29
580  FOR T=1 TO MX
590  TZ=T+MY:IF TZ>12 THEN TZ=TZ-12
600  DC=DC+MD(TZ)
610  NEXT T:MD(2)=29:DC=DC+MD(MO)-DA
620  PRINT:PRINT:PRINT
630  PRINT "SUBJECT HAS BEEN ALIVE"
640  PRINT DC: "DAYS AT START OF"
650  PRINT "MONTH OF INTEREST"
660  ' CALCULATE BEGINNING BIORHYTHMS
670  ' PP = PHYSICAL, MM = MENTAL
680  ' EE = EMOTIONAL
690  PX=DC/23:PY=23*INT(PX):PP=DC-PY
700  MX=DC/33:MY=33*INT(MX):MM=DC-MY
710  EX=DC/28:EY=28*INT(EX):EE=DC-EY
720  ' PRINT OUT BIORHYTHM CHART
730  ' FOR MONTH OF INTEREST
740  PRINT:PRINT "PRESS <RETURN>"
750  INPUT Q$
760  CLS
770  PRINT:PRINT:PRINT:PRINT
780  PRINT "BIRTH DATE"
790  PRINT " ";MO;"/";DA;"/";YR-1900
800  PRINT:PRINT "BIORHYTHM CHART FOR";MB;",";YB
810  MD(2)=28:LP=YB/4:IF LP=INT(LP) THEN MD(2)=29
820  XX=MD(MB)
830  PF=1:IF PP=23 THEN PF=-1
840  EF=1:IF EE=28 THEN EF=-1
850  MF=1:IF MM=33 THEN MF=-1
860  FOR TT=1 TO XX
870  PR=INT(PP-(23/2))
880  ER=INT(EE-(28/2))
890  MR=INT(MM-(33/2))
900  PRINT TT;" P = ";PR;" E = ";ER;" I = ":MR
910  ' UPDATE VALUES
920  PP=PP+PF:IF PP>22 THEN PF=-1
930  IF PP<2 THEN PF=1
940  EE=EE+EF:IF EE>27 THEN EF=-1
950  IF EE<2 THEN EF=1
960  MM=MM+MF:IF MM>32 THEN MF=-1
970  IF MM<2 THEN MF=1
980  NEXT TT
990  PRINT:PRINT:PRINT
1000  PRINT "ENTER 1 TO TRY NEW BIORHYTHM"
1010  PRINT "MONTH, 2 TO TRY NEW BIRTH DATE"
1020  PRINT "OR 3 TO END"
1030  INPUT QQ
1040  IF QQ=1 GOTO 350
```

Table 6-1 Continued

```
1050  IF QQ=2 GOTO 170
1060  END
1070  '
1080  '
1090  ' DATA FOR DAYS OF MONTH
1100  DATA 31,29,31,30,31,30
1110  DATA 31,31,30,31,30,31
1120  '
1130  '
1140  ' ROUTINE TO CHECK FOR LEAP
1150  ' YEAR FOR FEB. 29
1160  ' ROUTINE TO CHECK FOR LEAP
1170  PRINTYT=YR/4:YZ=INT(YT)
1180  IF YZ=YT GOTO 310
1190  GOTO 270
```

Finally, the date of the month is entered in lines 270 and 280 (DA). The birth date must not exceed the maximum number of days in the selected month, as stored in the appropriate array position (MD(MO)). February 29 presents a special case because this date can only occur during leap years, which are always evenly divisible by four. Line 300 calls up a validity checking routine (lines 1070 through 1090) in case this particularly "troublesome" date happens to be selected.

Now that the complete birth date has been entered, we need to tell the program what dates we are interested in for the biorhythm values. As written here, the program is set up to print out a full month of biorhythm values each time it is used. The month displayed must begin at the first of that month and run to the maximum date as stored in the MD array.

The year (YB) and month (MB) of interest are entered in lines 350 through 420. This section of the program is similar to the year and month portions of the birth date entry process. The one addition here is that the year of interest (YB) must be later (a larger value) than the birth year (YR). It is meaningless to try to calculate biorhythms for before the subject was born.

The program now has all the information it needs to start calculating the actual biorhythm values. Line 450 displays a short message to this effect. Lines 470 and 480 calculate the difference in months between the birth month and the biorhythm month. This accounts for any fractional portion of a year. This difference is called "MX" by the program.

The program then calculates how many days the subject has been alive at the beginning of the month of interest. This value is identified by the variable "DC" for "day count." First, it finds the days for the full years lived. This is found by multiplying the difference of YB – YR (called "YX") by 365.25 and taking the integer (INT) of this product. Why .25? This is a "fudge factor" to account for leap years. For every four years, the .25s will add up to one full day, just as the calendar indicates.

Lines 560 through 610 calculate the fraction of a year involved. Line 570 sets MD(2) (the February day count) to 28, unless the year of interest is a leap year. This array position is set back to 29 in line 610. This line also calculates the remaining days in the birth month (MO) after the birth date (DA).

Lines 620 through 650 display how many days the subject has been alive as of the first day of the month of interest. Line 690 sets PP to equal the current physical cycle value. Line 700 does the same thing for the intellectual cycle (MM) and line 710 takes care of the emotional cycle (EE). The user is prompted to press the "RETURN" key for the program to continue. On some computers this key might be marked "ENTER," or it might be marked with a reversing arrow or some other symbol.

Lines 770 through 980 display the physical (PP), emotional (EE), and intellectual (MM) values for each date of the month. On the positive half-cycles, one is added to each of these values. If the value exceeds the maximum cycle value, the flag is changed (PF, EF, or MF, as appropriate) and one is subtracted for each date until a value less than 1 is reached, when the appropriate flag is reset. This section of the program can be adapted to print out a hard copy (through your printer). Just change the print statements in lines 780, 790, 800, and 900.

When the complete month has been displayed, the user is given the option of trying a new biorhythm month, a new birth date, or ending the program. If one of the first two options is selected, the program cycles back to the appropriate spot to accept the new values to be entered. A typical output chart for this program appears as Table 6-2.

There are many ways this simple biorhythm program can be streamlined and improved, but this should be enough to help you get started.

Biorhythm clock

Of course, this chapter would not be complete without at least one electronic project involving biorhythms. The remainder of

Table 6-2 Typical biorhythm chart.

Birth date: 4/17/54
Biorhythm chart for 11, 1991

1	P = −6	E = −4	I = 14
2	P = −5	E = −3	I = 15
3	P = −4	E = −2	I = 16
4	P = −3	E = −1	I = 15
5	P = −2	E = 0	I = 14
6	P = −1	E = 1	I = 13
7	P = 0	E = 2	I = 12
8	P = 1	E = 3	I = 11
9	P = 2	E = 4	I = 10
10	P = 3	E = 5	I = 9
11	P = 4	E = 6	I = 8
12	P = 5	E = 7	I = 7
13	P = 6	E = 8	I = 6
14	P = 7	E = 9	I = 5
15	P = 8	E = 10	I = 4
16	P = 9	E = 11	I = 3
17	P = 10	E = 12	I = 2
18	P = 11	E = 13	I = 1
19	P = 10	E = 14	I = 0
20	P = 9	E = 13	I = −1
21	P = 8	E = 12	I = −2
22	P = 7	E = 11	I = −3
23	P = 6	E = 10	I = −4
24	P = 5	E = 9	I = −5
25	P = 4	E = 8	I = −6
26	P = 3	E = 7	I = −7
27	P = 2	E = 6	I = −8
28	P = 1	E = 5	I = −9
29	P = 0	E = 4	I = −10
30	P = −1	E = 3	I = −11

this chapter is devoted to a biorhythm clock, which can be set up to keep daily track of your physical, emotional, or intellectual cycle, or even all three. You can choose what you want to include.

This is actually one of the more extensive projects in this book. The schematic is broken up into four sections in Figs. 6-4 through 6-7. The last three sections are quite similar to one another. The complete parts list for this biorhythm clock project is given in Table 6-3.

Keeping track of biorhythms is not difficult, it's just a matter of counting. An up/down counter is needed because the biorhythm value goes up from a minimum value to a maximum value, then back down again. Each of the three biorhythm cycles has a different number of steps.

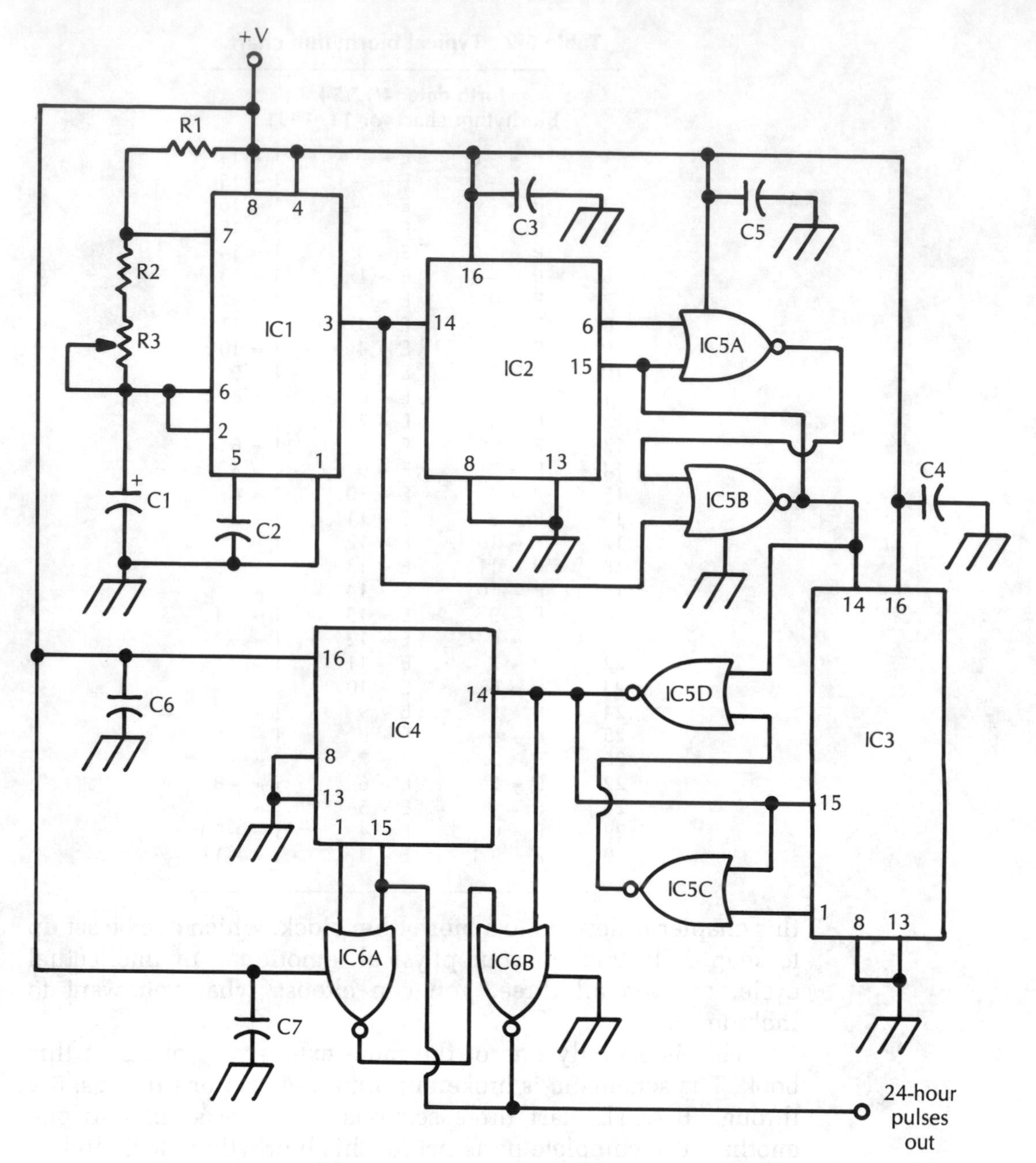

Fig. 6-4 *The 24-hour clock section of the biorhythm clock project.*

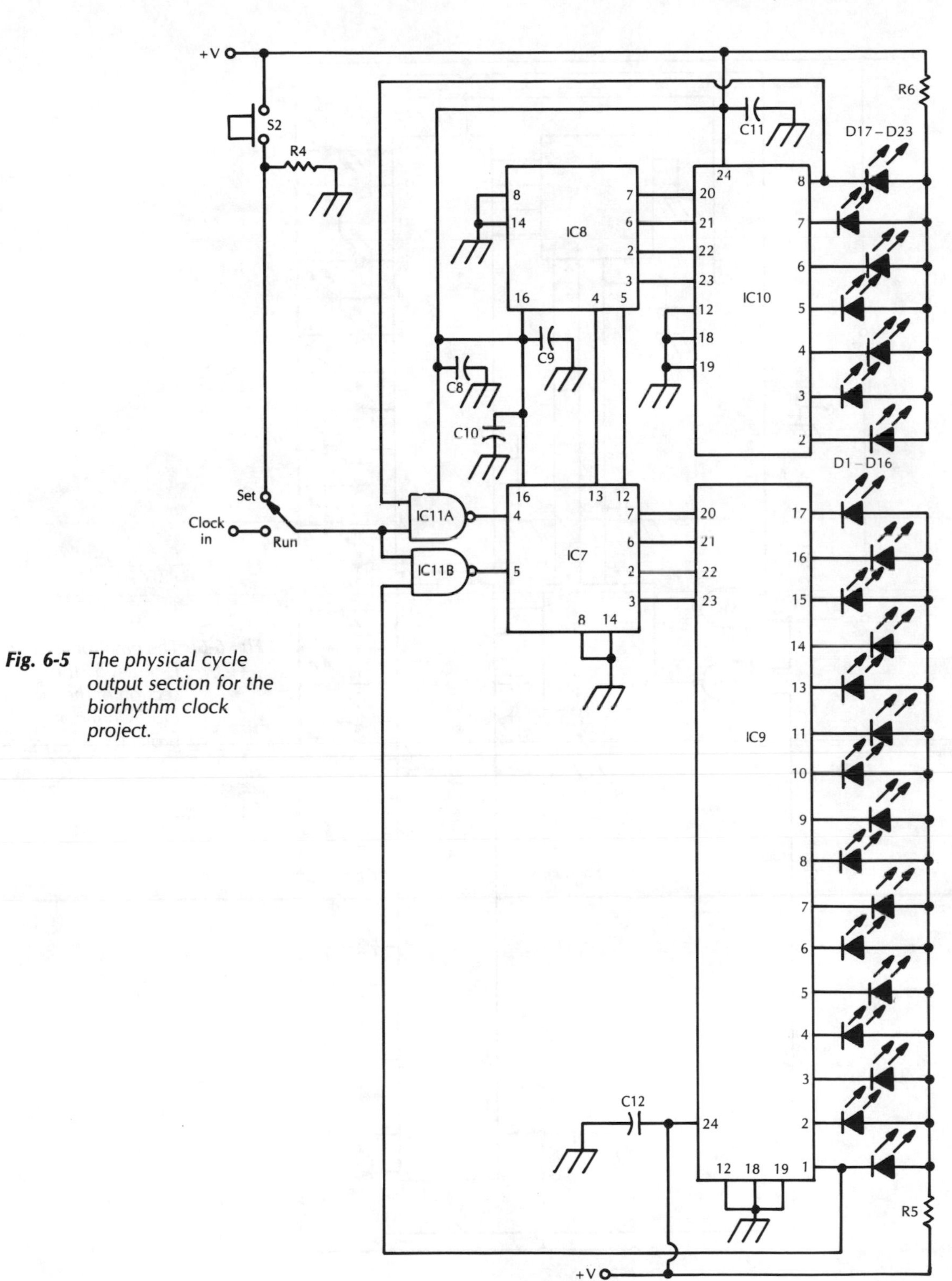

Fig. 6-5 *The physical cycle output section for the biorhythm clock project.*

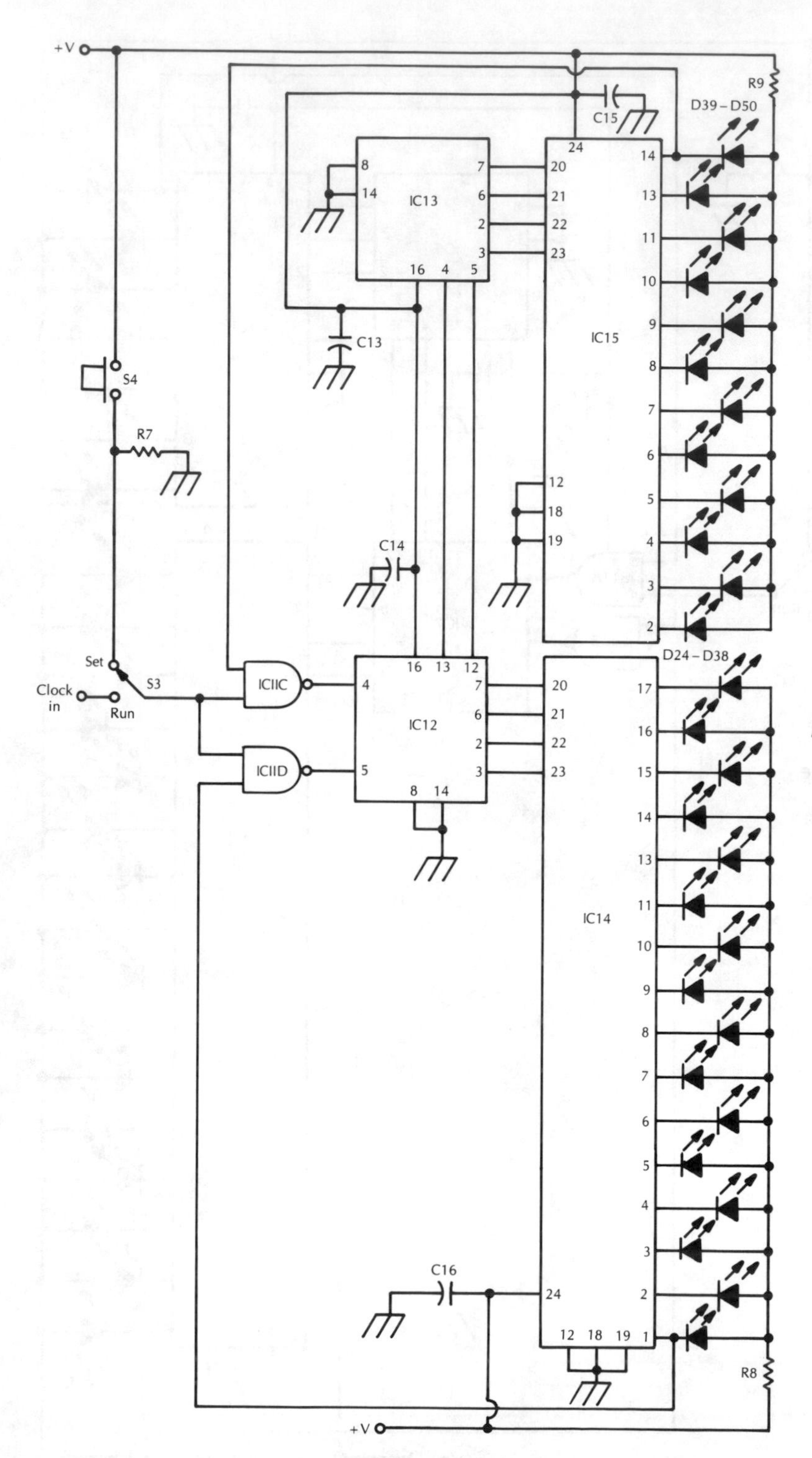

Fig. 6-6 *The emotional cycle is timed by this section of the biorhythm clock project.*

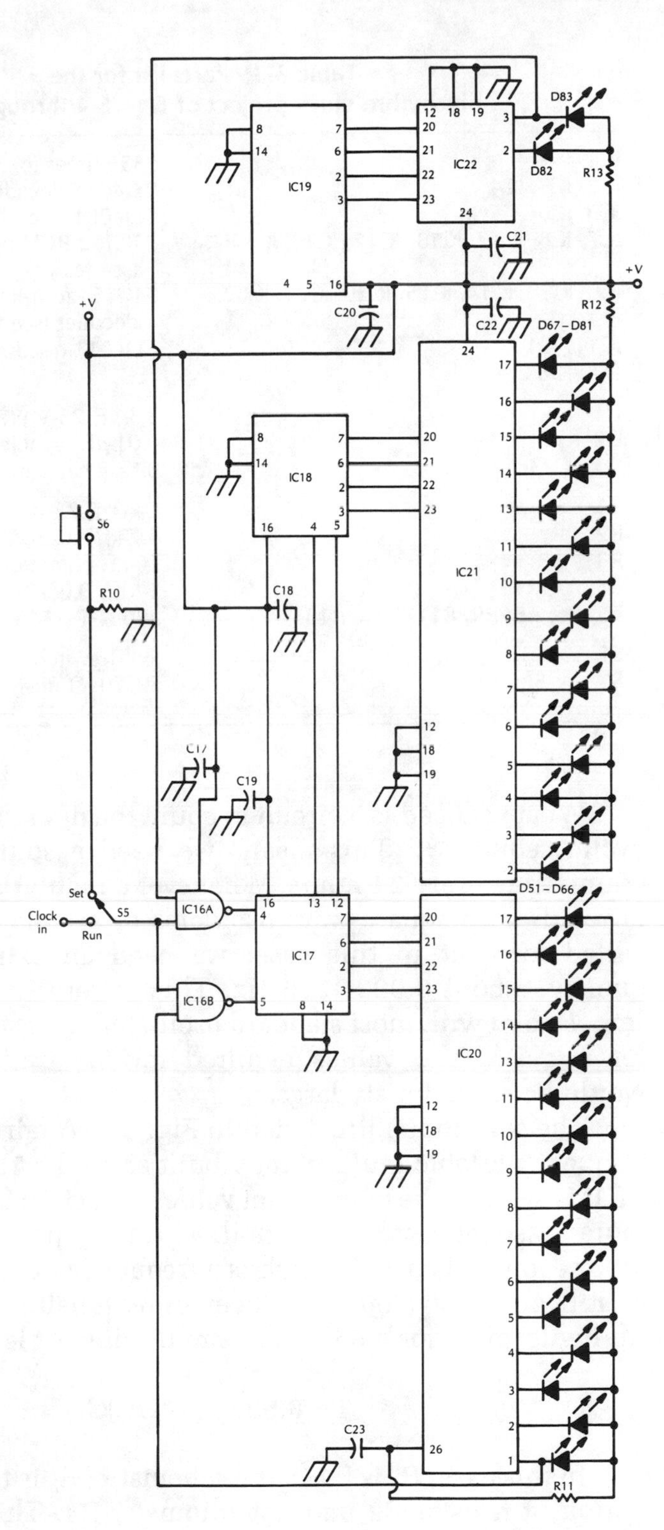

Fig. 6-7 *The intellectual cycle output section for the biorhythm clock project.*

Table 6-3 Parts list for the biorhythm clock project of Figs. 6-4 through 6-7.

IC1	7555 timer (or 555)
IC2, IC3, IC4	CD4017 decade counter/divider
IC5, IC6	CD4001 quad NOR gate
IC7, IC8, IC12, IC13, IC17, IC18, IC19	74C192 BCD up/down counter (see text)
IC9, IC10, IC14, IC15, IC20, IC21, IC22	74C154 four-line to sixteen-line decoder (see text)
IC11, IC16	CD4011 quad NAND gate
D1 – D83	LED
C1	3.3-μF, 25-V electrolytic capacitor
C2	0.01-μF capacitor
C3 – C23	0.1-μF capacitor
R1	1-kΩ, 0.25-W, 5% resistor
R2	4.7-kΩ, 0.25-W, 5% resistor
R3	10-kΩ potentiometer
R4, R7, R10	1-MΩ, 0.25-W, 5% resistor
R5, R6, R8, R9, R11, R12, R13	390-Ω, 0.25-W, 5% resistor
S1, S3, S5	SPDT switch
S2, S4, S6	SPST NO push-button switch

What we need is a circuit to count the days in the appropriate cyclic sequences. This means we need a source of pulses to count—one every 24 hours. Whenever a regular stream of pulses is required, a square-wave oscillator or astable multivibrator is called for. But in this case, we need an extremely low frequency—about 0.000 011 5 Hz. This is very impractical (if not impossible) with most standard oscillator and astable multivibrator circuits. The values required for the timing components would be ridiculously large.

The solution is illustrated in Fig. 6-4. A fairly standard low-frequency astable multivibrator built around a 7555 or 555 timer (IC1) is set up. The component values are set up for a long, but far more reasonable timing period. A timing period of 5 minutes (300 seconds) is used, which is a frequency of 0.003 33 Hz. This is still quite low, but no longer unreasonable. Remember, the timing formula for a 555 astable multivibrator is

$$T = 0.693(R_a + 2R_b)C_1$$

In this case, R_a is R1 in the schematic; R_b is the series combination of resistor R2 and potentiometer R3. This potentiometer allows you to fine-tune the exact timing of the multivibrator. It takes a little patience, but check out the timing cycle with a voltmeter and a stop watch until it is as close to 5 minutes as possible. Any inaccuracy in this timing period will be magnified considerably by the later frequency divider stages.

The nominal setting for potentiometer R3 is about 160 kΩ, so

$$\begin{aligned} R_b &= R_2 + R_3 \\ &= 470\,000 + 160\,000 \\ &= 630\,000\ \Omega \\ &= 630\ \text{k}\Omega \end{aligned}$$

The other timing component values are

- R1 = 470 kΩ (470 000 Ω)
- C1 = 250 μF (0.000 25 F)

This gives our astable multivibrator a nominal timing period of

$$\begin{aligned} T &= 0.693(R_a + 2R_b)C_1 \\ &= 0.693(470\,000 + (2 \times 630\,000)0.000\,25 \\ &= 0.693(470\,000 + 1\,260\,000)0.000\,25 \\ &= 0.693 \times 1\,730\,000 \times 0.000\,25 \\ &= 299.7 \text{ seconds} \\ &= 4.99 \text{ minutes} \end{aligned}$$

Of course, the exact setting of potentiometer R3 will correct for the small error here.

It is advisable to use a screwdriver-set trimpot for R3. Adjust the control properly, then leave it alone. A small drop of glue or paint will hold the turn shaft in place.

Now that you have 5-minute pulses, the next step is to use some counter stages to serve as a frequency divider. IC2 divides the pulses by 8, so this stage has 1 output pulse every 5 × 8 minutes, or once every 40 minutes. Next, IC3 further divides the pulse rate by 6, so the timing is now 1 pulse every 6 × 40 minutes, or once every 240 minutes. Next, IC5 divides the pulses by another 6, giving us 1 pulse every 6 × 240 minutes or once every 1 440 minutes. There are 1 440 minutes in a 24-hour day (1 440/60 = 24), so we have finally achieved our goal for this first stage of our biorhythm clock. The output from this section of circuitry is 1 pulse every 24 hours. Now we can begin counting the actual biorhythm cycles.

This 24-hour pulse generator can be used in many other projects. It is especially helpful in automation projects that trigger some action on a daily basis.

Figure 6-5 shows the circuitry for the physical cycle portion of our biorhythm clock. IC6 and IC7 are cascaded up/down BCD counters. The count direction is controlled by IC6C and IC6D (sections A and B of IC6 are used in Fig. 6-6). The BCD count values are fed to IC9 and IC10, a pair of 4- to 16-line decoder chips. Only seven of the output lines of IC10 are used. Coupled with the 16 output lines of IC9 this gives us 16 + 7 or 23 possible outputs. There are 23 daily steps in the physical biorhythm cycle. When the maximum count (23) is reached, the counters are automatically reversed to down counting. When the minimum count (1) is reached, the direction is changed again for up counting.

For each day, one and only one LED is lit, indicating the physical biorhythmic value for that day. Diode D13 is the zero (critical day) indicator. This LED should be specially placed, or perhaps a larger or different colored LED can be used to mark this important point in the cycle. Diodes D14 through D23 are for positive physical values. Diodes D12 down to D1 are for negative physical values.

When you first start using the biorhythm clock, you must set it to the proper biorhythm value (which can be found with the program presented earlier in this chapter). This is done by moving switch S1 to the "SET" position. This disconnects the 24-hour pulse generator and allows you to manually enter pulses via push-button switch S2. A good quality bounce-free push-button switch is strongly recommended for this application. When this switch is open (not depressed), the signal is held low through resistor R4. Pushing the switch closed pulls the signal high, entering one pulse to be counted.

Figure 6-6 shows the very similar circuitry for the emotional cycle, which runs for 28 days. There are five more LED outputs in this section because of the longer biorhythm cycle. Because the emotional cycle is made up of an even number of days (28), there is no zero position. The critical day is midway between day 1 (D36) and day − 1 (D35). Higher-numbered LEDs are for the positive portion of the emotional cycle, while lower-numbered LEDs are for the negative half of the cycle. Except for the slightly longer cycle (number of output LEDs), this circuit is identical to the physical cycle circuitry shown in Fig. 6-5.

Finally, we come to the intellectual cycle, shown in Fig. 6-7.

Because the intellectual cycle lasts 33 days, a third counter/ decoder stage is required. Otherwise, it is just repetition of the same basic circuit used for the other biorhythm cycles. The zero (critical) day is represented by LED D67.

You might have difficulty finding the CMOS 74C192 and 74C154 ICs. You can use TTL devices (74192 and 74154). Add the circuit shown in Fig. 6-8 to interface the TTL circuitry (Figs. 6-5, 6-6, and 6-7) to the CMOS circuitry of Fig. 6-4. The TTL circuitry must be operated from a tightly regulated +5-V power supply. CMOS devices can handle larger supply voltages. This optional interface circuit matches the signal levels so the CMOS and TTL circuits can "talk" to each other. A parts list for the TTL to CMOS converter circuit is given in Table 6-4.

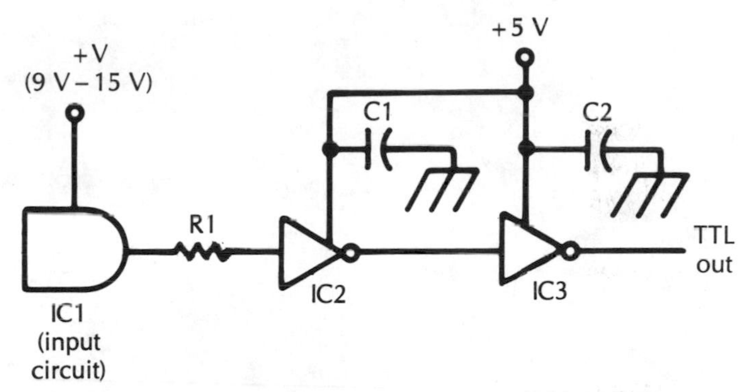

Fig. 6-8 *This circuit can be used to interface a TTL output with a CMOS input.*

Table 6-4 Parts list for the TTL to CMOS converter circuit of Fig. 6-8.

IC1	CMOS gate (depending on circuit)
IC2	CD4049 CMOS inverter (or CD4050)
IC3	7404 TTL inverter
C1, C2	0.1-μF capacitor
R1	1-kΩ, 0.25-W, 5% resistor

7 Kirlian photography

THE WORD "PHOTOGRAPHY" MEANS "WRITING WITH LIGHT"—"photo" means "light" and "graph" means "write." In ordinary photography, you make a permanent record of the visible light patterns that existed before the camera's lens at the instant the photograph was taken.

Nonstandard photography

While snapshots are, by far, the most common type of photography, other types also exist. In some specialized forms of photography, nonvisible light is used. The resulting photograph shows things that are not directly visible to the naked eye.

A good example of this is the X-ray photograph. Who hasn't had an X ray taken by the doctor or dentist? This is just a modified form of ordinary photography. In ordinary photography, the film is sensitive to visible light, just like the human eye. Where bright light strikes the film, a light spot shows up on the finished photograph; where less light strikes the film, a shadow or dark spot shows up on the processed photograph. An X-ray camera, on the other hand, is insensitive to visible light. This type of camera uses a special film that can only "see" a special type of invisible light, known as an X ray. (When X rays were first discovered, scientists knew they were dealing with some sort of ray, but its exact nature was unknown, so they started referring to these "mystery" rays as "X rays." "X" is often used in mathematics and science to represent an unknown item. Even when more was learned about X rays, the name remained.)

Ordinary visible light is relatively weak and can be blocked by almost any object. X rays, however, can pass through soft, porous materials, such as skin, but they are blocked by hard materials, such as bone. So when an X ray is taken of a person, the flesh doesn't show up, and the bones appear as dark shadows.

Photographs can be taken with almost any type of light, which might or might not be directly visible to the human eye. You might be wondering just how different types of light are different. After all, light is light, isn't it? Either an area is light or it is dark. Right?

Actually, light, like sound, can occur at different frequencies over a wide range. Visible light accounts for only a narrow band of the entire spectrum of possible light frequencies. Within the visible range, the frequency of the light waves determines the color we see. For example, blue light has a higher frequency than red light. White light is comprised of all visible light frequencies.

Many types of light have frequencies too high or too low for the human eye to detect. For example, X rays have frequencies considerably higher than the uppermost visible frequency. Infrared light is just below the lowest visible frequency (red). Similarly, ultraviolet light is just above the highest visible frequency (violet). A number of specialized photography techniques have been devised to work with different light frequencies to make the invisible visible.

Kirlian photography

We are concerned with a special type of photography known as Kirlian photography (pronounced "keer-lee-an"). This odd sounding name doesn't have any special meaning or technical derivation. Kirlian photography was simply named after the man who developed the technique. A relatively obscure scientist named Semyon Kirlian invented what became known as Kirlian photography in 1939. It is still a fairly unfamiliar and highly controversial area. Kirlian photography is a form of electrophotography. As this name suggests, electricity is used to take this type of photograph.

In a Kirlian photograph, objects are surrounded by colorful auras. Some claim these auras can be used to detect diseases and other problems in plants and animals, even before external symptoms appear. New Agers often read even more meaning into the auras of Kirlian photograhs. It is a common New Age belief that people are surrounded by auras, which indicate their inner state on the physical, mental, and spiritual levels. Many psychics claim to be able to see and read the auras around a person. With a Kirlian photograph, anyone can see the auras. (Of course interpreting what is seen is a whole different matter.)

Like X-ray photography, Kirlian photography does not use an actual lens or camera (at least, not in the usual sense). The resulting image is not really a recording of light patterns. Instead of ordinary light, Kirlian photography uses a high-frequency, high-voltage source to produce the images on the photographic film.

At first glance, the whole idea of Kirlian photography might seem somewhat frivolous. Yes, the colorful auras in a Kirlian photograph look nifty, but so what? Why would anyone want to bother? Well, Kirlian photography has been the subject of some serious research, even by the U.S. government, primarily at the Biochemistry Laboratory at the Naval Air Development Center. This unit, located in Warminster, Pennsylvania, was once at the forefront of Kirlian photography research, but it has since been disbanded, although research in this field continues at other important laboratories.

Auras

In a Kirlian photograph, persons and objects are surrounded by brightly colored envelopes of light. These envelopes of light are generally known as auras or coronas. Traditional scientists seem to prefer the term "corona." I guess "corona" sounds more "scientific" and "aura" sounds more "mystical." Ultimately, both words refer to the same thing. I prefer the word "aura" simply because it is more descriptive to the layman. I think it is easier for most of us to visualize an aura than a corona.

Often these Kirlian auras bear a strong resemblance to the halos in old religious paintings. Some in New Age circles argue that everyone is always surrounded by an aura, although it is usually invisible (at least to most eyes), and that the halos in these old artworks represent incompletely perceived auras. Most researchers in this field, however, would probably say there is no real connection between the halos of early art and the auras of Kirlian photography beyond the purely coincidental resemblance in appearance.

There is no question that auras appear in Kirlian photographs. They're right there on the processed picture for everyone to see. But there is a great deal of debate and controversy over just what these mysterious auras really are and what they might mean. No one yet knows for sure just what causes these auras to appear in Kirlian photographs. Of course, there are many theories about these Kirlian auras. Some of the theories are fairly plausible, while others seem quite far-out. In the following discussion I will leave it to the individual reader to decide which (if any) of the theories makes the most sense.

Some New Agers believe that such auras are direct visual contact with the spiritual realm. Different colors represent different spiritual attributes or conditions. For example, a person with a lot of gold or white in his aura is thought to be highly advanced spiritually, while a black aura indicates very severe spirtual problems or impending death.

Kirlian photographs of inanimate objects also show auras. The auras surrounding inorganic objects (such as rocks) are generally not as dramatic as those surrounding living creatures or plants. When a plant or animal dies its aura is also decreased. But even dead and inorganic subjects still have definite auras in Kirlian photographs. This is accounted for by the belief that everything in the material realm is a manifestation of activity within the spirtual realm. On a purely spirtual level, everything is alive, including apparently inanimate objects.

A theory that sits more comfortably with people trained in traditional science and physics suggests that all physical substances are surrounded by a matter-energy field, or plasma. This is particularly true of living organic matter. Recent research in physics (especially quantum physics) has revealed that energy and matter are not inherently different. In fact, matter appears to be a specialized form of contained energy.

Ironically, at the time Semyon Kirlian was doing his pioneering work in this field, this type of explanation would have been considered totally nonsensical, if not completely incomprehensible. Even 20 or 30 years ago, the whole idea would have been pooh-poohed by most scientists as far-out fantasy. But modern experiments in quantum physics makes the matter-energy field theory quite plausible after all.

Actually, the "New Age" spiritual theory and the "scientific" plasma theory aren't really in as much competition as it might seem at first glance. These two explanations of the aura phenomenon in Kirlian photography might ultimately be saying the same thing, just using different language and imagery. If the aura is a manifestation of spiritual essence, why shouldn't it take the form of a matter-energy plasma? Are the two theories really so different, or are they describing the same thing from very different angles?

Of course, the aura formations might be due to chemical reactions or some other physical phenomenon without any spiritual, or even quantum, component. But how can we scientifically rule out the possibility of a spiritual connection? At present the true scientific position is, "No one knows."

Possible applications

The auras in Kirlian photographs, whatever they might ultimately be, do seem to have some relationship to the (living) subject's well-being. The aura around an inanimate object appears to be fairly constant, but if Kirlian photographs are taken of a living subject at different times, there are fluctuations and sometimes dramatic differences in the photographic aura. Kirlian photographs and their auras have been used experimentally in medical diagnosis, and thus far this appears to be a valid and potentially valuable area of research. Alcohol or drug consumption and stress can apparently affect the appearance of the aura in a Kirlian photograph.

One of the studies sponsored by the U.S. government measured the diameter of the aura (or corona) at the fingertip of a human subject. There were statistically significant differences detected in this study. The aura size seemed related to stress. Moreover, physical stress and emotional stress seemed to have opposite effects on the subject's aura. Physical stress (induced by vigorous exercise) resulted in a larger than normal aura, while emotional stress (resulting from fatigue or similar conditions) caused smaller than normal auras. A nonstressed subject had an aura of moderate size.

There also appears to be evidence that certain diseases and medical conditions can reduce the size of a subject's fingertip aura. Despite a number of studies, using Kirlian photography for diagnosis along these lines is far from foolproof, at least at present. The results appear to be statistically significant—that is, they are right more often than not—but there are still an unacceptably high number of "misses." Further research might refine the techniques involved for more precise and reliable results.

For nonorganic subjects, Kirlian photography has been used to test materials, mostly in engineering applications. Unlike most other commonly used types of material testing, Kirlian photography is nondestructive. The material being tested is not damaged or destroyed in the process. Such techniques appear to hold great promise for future industrial applications.

Keep in mind, however, that at this time, Kirlian photography is still considered to be an experimental technique. The aural effects described here have not been conclusively proven. Some researchers feel the aural patterns appearing in Kirlian photographs are random or are dependent more on the photographic conditions (the specific voltage or frequency used) than on any condition within the subject being photographed.

Overall, the experimental results thus far have been highly provocative, but more research is required before we can place much faith in these techniques. There are a great many experimental parameters and controls that must be carefully considered and accounted for before it is scientifically reasonable to draw any definite conclusions about Kirlian photographs and their meanings (if any).

The phantom leaf effect

Perhaps one of the most intriguing, mysterious, and fascinating phenomena associated with Kirlian photography is the ''phantom leaf effect.'' If a small part of a leaf is cut off (about 2% to 10% of its total surface area), and a Kirlian photograph is taken of the leaf, the resulting aura will sometimes look as if the entire leaf is present. There will be no notch or hole in the aura. Apparently the energy pattern of the leaf is still whole, even though the physical leaf is incomplete. The phantom leaf effect does not always occur. In fact, it is rather rare. But when it does occur, it is unquestionably impressive. It has occurred frequently enough to be considered a real phenomonon and not just a fluke.

No one can yet explain the phantom leaf effect. Some New Agers hold that it is proof that on the spiritual level, all things are whole. Even though the physical body is damaged or destroyed, the spiritual body cannot be harmed. The problem with this explanation is that it doesn't explain why the phantom leaf effect doesn't occur every time. Still, it is a highly intriguing idea.

While it is exceedingly upsetting to skeptics, it seems that there really is something going on in Kirlian photography that is

beyond our present level of understanding. This is not to say that it is in any way magical or even nonphysical, or that it can never be understood scientifically—just that science hasn't solved the puzzle of Kirlian photography yet.

More on diagnostic aura research

Earlier I mentioned the study run by the U.S. government that involved measuring the Kirlian auras around the fingertips of military personnel. Kirlian photographs were taken of the subject's fingertips and the resulting auras were carefully measured and analyzed. There was a statistically significant (though not absolute) link between the size of the fingertip aura and the subject's stress or ease.

If the subject exercised vigorously to create physical stress before the Kirlian photograph was taken, the resulting aura was larger in diameter than the overall test average. Conversely, if the subject was fatigued or suffering from some similar mental or emotional stress, the Kirlian photograph showed a smaller diameter aura than the overall test average. These results were consistent, but not invariable. The diagnostic readings were far from foolproof, especially when any sort of diagnosis was attempted from just a single Kirlian photograph of a given subject. Apparently there is some difference between individuals in the normal aura diameter. It seems likely that for Kirlian photography to be a useful diagnostic tool, the diagnostician needs a reference Kirlian photograph of the individual being examined. This reference photo should be taken when the subject is experiencing a minimum of both mental and physical stress.

One study compared the fingertip aura diameter of 120 adult human subjects. While most unstressed subjects had fingertip auras of more or less the same size, about 20% of the subjects in this study had significantly smaller than average fingertip auras. One possible explanation was suggested by later events when it was discovered that about half of these "small aura" subjects were suffering from some sort of medical problem at the time of the study. In some cases, the ailment had not been noticed or diagnosed before the Kirlian photographs were taken.

It is tempting to infer that the smaller than average fingertip auras were due to the unrecognized (at the time) medical problems. But half of those subjects with smaller than average fingertip auras apparently did not have any medical problem to

account for their reduced auras. More importantly, this particular study made no report on the health conditions of the 80% of the subjects who had "normal" sized auras. Also, there is a question as to why the subjects with medical problems did not show the same aural changes as physical stress, but rather, the effects looked more like mental stress.

Another key study performed at the University of California by a psychologist named Dr. Thelma Moss combined Kirlian photography with another New Age phenomenon—laying on of hands healing. You could call this a form of "faith healing," although some additional implications are usually read into this second term.

Moss took Kirlian photographs of the fingertips of both the healer and the patient before and after the healing process and compared the resulting auras. Before the healing process, the healer's fingertip was generally bright and whole, while the patient's was weak and incomplete. After the healing process, this was reversed, the healer's fingertip aura was now noticeably weakened and broken (though not as badly as the patient's original fingertip aura), while the patient's fingertip now exhibited a full, bright aura. Follow-up Kirlian photographs showed that the healer's fingertip soon recovered its full original condition. This did not occur with control patients who were not subjected to the healing process.

Such studies are highly controversial, and many traditional scientists strongly contest the results. Still, there have been a number of separate studies indicating that there might well be something to New Age nonmedical healing. It might be due to some sort of psychosomatic, mind-over-matter effect. The link between emotional state and health or healing capabilities seems to be close to proved (but there is still some room for scientific doubt). On the other hand, the results from such unorthodox healing processes are still far from reliable or foolproof. But the fact that they sometimes work is striking in itself and calls for an explanation beyond a mere skeptical scoffing at the results as a "fluke" or "experimental error."

Studies of Kirlian photography, like the ones described here, are certainly provocative and suggest areas for future research projects, but they are often (as in this case) terribly incomplete. This is not to blame the researchers for sloppy work. Any given study can do just so much or it becomes unwieldy and unreliable in the number of variables to be controlled. The point here is that

we must avoid the strong temptation to read too much into the results. It's still too early in the game to draw any definite conclusions about Kirlian photography as a diagnostic tool.

The skeptics

A certain degree of healthy skepticism is advisable when looking into the field of Kirlian photography, as with all of the topics covered in this book. No theory has been proven yet. Many New Agers consider Kirlian auras "proof" of spiritual influences beyond the material realm. On the other hand, when a scientifically oriented magazine ran an article on Kirlian photography a few years ago, one reader indignantly wrote in, accusing the magazine of "having done a disservice to your readership and to science by stating, albeit obliquely, that there could be something beyond the realm of known physics that influences the Kirlian 'aura.' " Probably a number of other readers felt the same way.

This irate "scientific" believer cited an article in the Spring 1986 issue of *The Skeptical Inquirer* ("A Study of the Kirlian Effect." by Arleen J. Watkins and William S. Bickel) that described possible variables (such as pressure, temperature, conductive residues, exposure time, and so forth) that could possibly create the Kirlian images without, in the words of the letter writer, "resorting to the paranormal."

As I have tried to explain throughout this book, any paranormal effect is only paranormal because it is inexplicable by the current level of science. Electricity was once considered a paranormal phenemonon (although different words were used—it was usually called "magic.") There are definitely lots of things beyond the realm of known physics. This is why theoretical research in physics continues. It is arrogant, unscientific, and rather stupid to believe that modern science has found all the answers.

If Kirlian photographs do work as claimed by some believers, that doesn't necessarily mean that anything paranormal or occult is involved, any more than the functioning of a transistor is a paranormal or occult phenomenon. The matter-energy field theory, for example, suggests that Kirlian auras might be related to the odd phenomena observed in quantum physics. We aren't necessarily dealing with something paranormal or magical, just something that is at present unknown.

The parameters suggested in *The Skeptical Inquirer* article might account fully or partially for the effects of Kirlian photography, or they might not. Of course, for their explanation to be taken seriously, the skeptics need to account for the apparent statistical links with other phenomena (such as stress) that show up in Kirlian photography experiments.

Clearly, more research is needed to test the effects of each of these (and many other) variables. For the true scientist, the questions raised by Kirlian photography are far from answered. Kirlian photography might be nothing more than an amusing illusion (although the statistical consistency of the studies so far seems to make this a bit unlikely). It might be a phenomenon resulting from known physical principles in complex, and as yet unknown ways, or it might be caused by previously unknown physical principles of some kind. It might even be a totally nonmaterial, spiritual ("paranormal," if you must) phenomenon. The fact is, no one knows, yet.

If you find the questions raised by Kirlian photography intriguing (regardless of what you believe), you might want to do some experimentation on your own. My goal here is to help you get started on your own experiments. You take it from there. Who knows? You might even be the person who finally finds the answer.

The experimental circuit

Regardless of what you might believe about the auras appearing in Kirlian photographs, you might want to experiment with this unusual technique on your own. To take Kirlian photographs, you need a high-frequency, high-voltage source. This requires an electronic circuit. A suitable circuit for simple experimental purposes is shown in Fig. 7-1. Naturally, this project is nowhere near as sophisticated as the equipment used in serious scientific laboratories, but it is quite sufficient to get you started taking your own Kirlian photographs. The suggested parts list for this project is given in Table 7-1.

Not counting the special coil unit (T2), which is discussed below, the cost for constructing this project should be no more than $50 to $60, using all new components. Of course, if you have a well-stocked electronics junk box, you might be able to reduce this cost considerably.

The exact transistor types used in this circuit are not critical, provided they can handle fairly high power levels. Notice that Q1

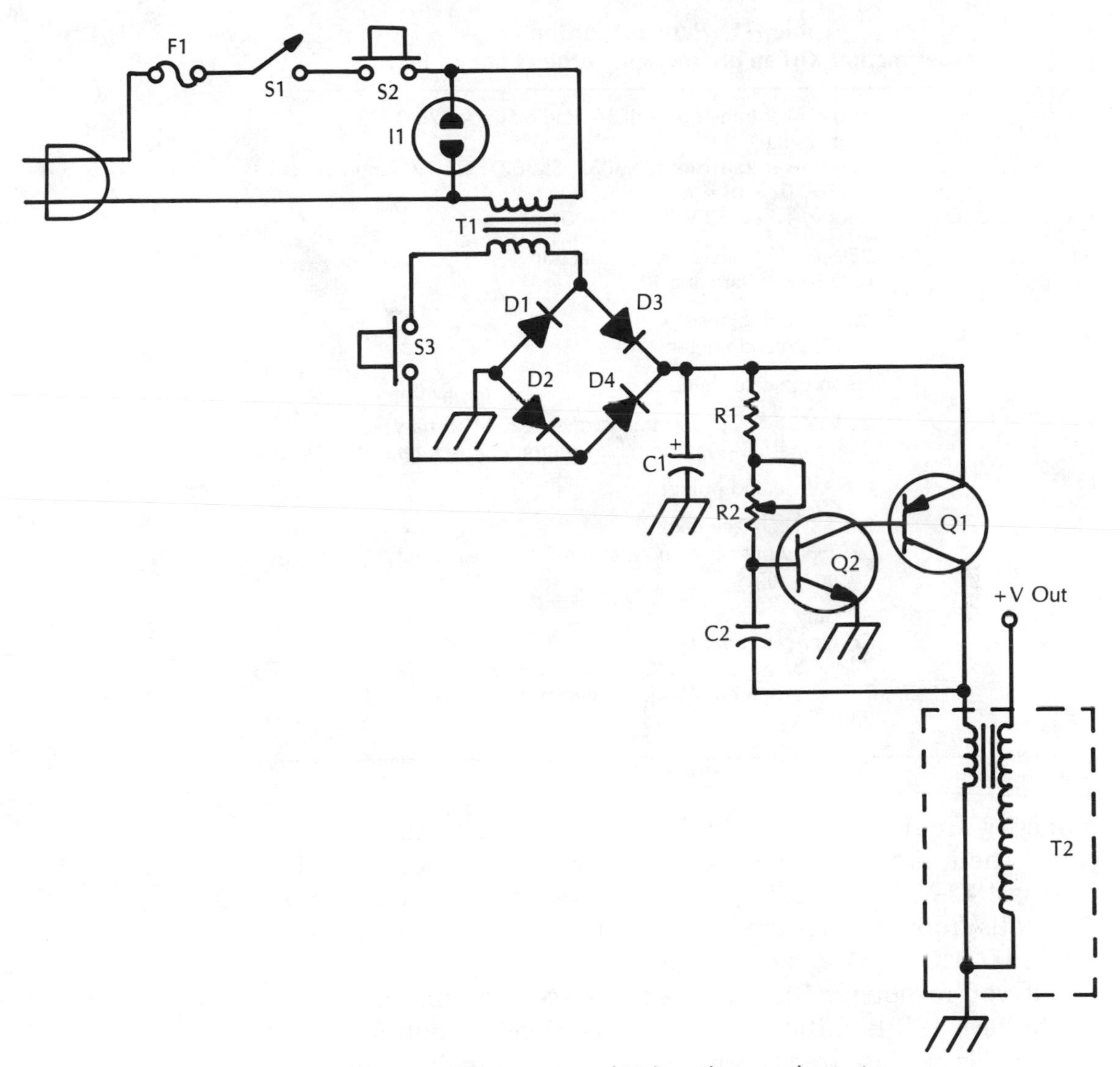

Fig. 7-1 The experimental Kirlian photography project.

is a pnp-type device, while Q2 is an npn unit. In use, these transistors can run very hot, so use large heat sinks on both of these components. It is not a good idea to operate this circuit without heat sinking the transistors. They will run very hot.

The only really unusual component in this project is T2. Actually, this is not a true transformer, but a standard 12-V, 3-terminal automotive ignition coil. You probably won't find this component at your electronics shop, but it should be available from an automotive supply store. You might be able to save some

Table 7-1 Parts list for the experimental Kirlian photography project of Fig. 7-1.

Q1	pnp power transistor (MJE34, Radio Shack 276-2027, or similar)
Q2	npn power transistor (2N3055, SK3027, GE-19, Radio Shack 276-2020, or similar)
D1, D2, D3, D4	Diode, 1.5-A, 50-V PIV (1N4001 or similar)
C1	250-μF, 35 V electrolytic capacitor
C2	0.025-μF mylar capacitor
R1	4.7-kΩ, 0.5-W resistor
R2	20-kΩ potentiometer
I1	Neon indicator lamp unit
T1	25-V, 450-mA power transformer (120-V primary)
T2	3-terminal, 12-V automotive ignition coil (see text)
F1	0.1-A fuse and holder
S1	SPST switch (see text)
S2, S3	SPST NO push-button switch (must be heavy duty) (see text)
	AC plug
	Copper-clad board
	Plastic bag
	Thin, transparent plastic (2 sheets)
	Photographic film

money by finding a suitable ignition coil at an automobile junkyard. There are no special requirements for T2, just use the cheapest 12-V, 3-terminal ignition coil you can find. Used ignition coils from an automotive junkyard can sometimes be found for under $10.

If you're not familiar with automotive ignition coils, you might have a little difficulty finding the high-V + output terminal. This is because in its normal use (in an automobile's engine) the connection is made with a heavy-gauge spark plug wire clip. The high-voltage terminal is inside a small tube-like projection at the top of the ignition coil. The stripped end of a heavy-gauge wire should be inserted into the opening of this projection. Make sure there is a good, strong mechanical connection here. Except for the ends, this wire should be well-insulated. Remember, it is carrying a dangerous level of electrical power. The free end of this wire is connected to the baseboard of the actual Kirlian photography assembly, which is discussed in the next section of this chapter.

It is extremely important to remember that this project is a high-voltage circuit, with the capability to put out very large current levels. Careless use of such a high-power circuit can result in a nasty electrical shock, possibly causing serious injury or even death. Be very, very careful when experimenting with this type of circuit, especially when taking Kirlian photographs of living subjects. Suitable precautions to use when taking Kirlian photographs are discussed in detail later in this chapter. Please read this information carefully and apply it at all times. Don't try to use shortcuts. Taking foolish risks could well have disastrous consequences.

When constructing the project, make sure that the ignition coil (T2) is well-insulated and can't be accidentally touched when power is applied to the circuit. Enclose this component in its own nonconductive chassis to prevent shock hazards. You can use a small plastic project case with appropriate holes drilled in it, or you can fashion a chassis from a short cardboard tube, using circular pieces of plastic or plexiglass to close off the ends of the tube.

Three connections must be made to the ignition coil—the collector of Q1, ground, and the high-voltage terminal. A well-insulated binding post should be used for this terminal. Mount the binding post directly onto the nonconductive coil chassis, with the wiring enclosed within the chassis. It is not a good idea to use a metal (conductive) chassis for this circuit, but if you do, make sure that the circuitry is completely insulated from the chassis. Be particularly careful with the ground connections.

Make sure you construct the project so that no one can touch any part of the circuit while power is applied. DO NOT USE AN EARTH GROUND in this circuit. The circuit must be entirely self-contained, with an internal ground only. If an earth ground is used, the shock and fire hazards are unacceptably high. Let me repeat, do not use an earth ground in this circuit!

Use good-quality, well-insulated push-button switches for S2 and S3. It is not a good idea to use cheap (and potentially unreliable) switches in this project. Power switch S1 can be omitted if you wish. It is included as a safety precaution against the possibility of the push-button switches being inadvertently closed before you are ready.

Do not under any circumstances consider omitting the fuse (F1). It is very inexpensive insurance against possible disaster. Remember, this is a voltage multiplier circuit. If something goes wrong, the current can quickly run away to excessively high and dangerous levels. The fuse blows if the current drawn goes too high, and this shuts the entire circuit down—hopefully before any serious damage is done. You can use a smaller fuse, if you like, but do not increase the fuse's rating above the 0.1 A specified in the parts list. Use only a fast-blow fuse in this application, not a slow-blow fuse.

It is not recommended that you use this circuit to take Kirlian photographs of living creatures or people. It can be done, but because of the high currents and voltages involved, it is very risky unless you know exactly what you are doing. It is probably best for most hobbyists to restrict their Kirlian photography experiments to inanimate subjects, such as small stones and other objects, or to plants. The procedure described in the next section of this chapter assumes a small, somewhat flat, nonliving subject.

Making the Kirlian photograph

To take a Kirlian photograph, you do not need a camera. The image is applied directly to the film. In addition to the circuit of Fig. 7-1, the subject to be photographed, and the film, you will need a piece of copper-clad board (an unetched PC board) and two pieces of thin, transparent plastic (0.01 inch is a good thickness for the plastic sheets used in this project). These elements are arranged in a sandwich configuration. The copper-clad board goes on the bottom with the film on top of it. One of the transparent plastic sheets is placed over the film. Then the subject to be photographed is placed on top of the plastic sheet and the second plastic sheet is placed on top of the subject. This arrangement is illustrated in Fig. 7-2. You can see why a small, flat object should be used as the subject here. In a full laboratory, much more sophisticated equipment and techniques are used, so the size and shape of the subject is far less limited.

The high-voltage V+ output from the generator circuit of Fig. 7-1 is connected to the copper-clad board at the base of our "sandwich." If the subject being photographed is not alive, best results can be obtained if it is electrically connected to the circuit's ground. Note that this is NOT earth ground! To use earth ground can be extremely dangerous. Never ground any living subject. A severe (and possibly fatal) shock is almost inevitable if a living subject is grounded. The resulting Kirlian photograph

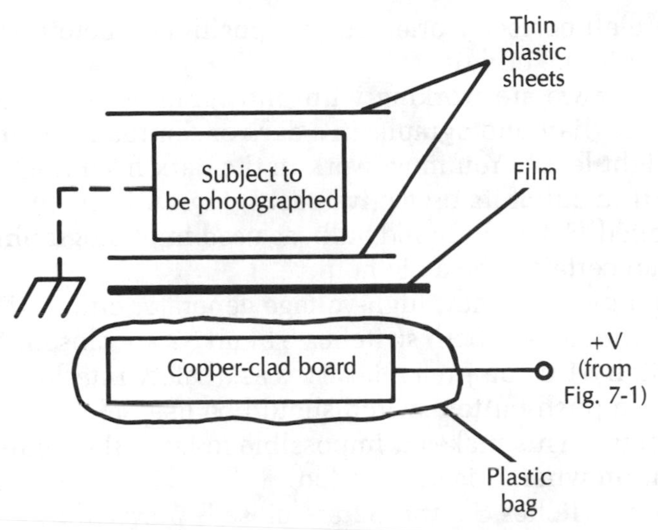

Fig. 7-2 *The required elements for making a Kirlian photograph are arranged in a sandwich-like assembly.*

won't be quite as clear and sharp if the subject is not grounded, but a somewhat better experimental photograph is never worth the risk involved in grounding a living creature or person. Do not do it under any circumstances!

The copper-clad board should be large enough to hold the subject (or a portion of the subject) being Kirlian photographed. About 5 inches square is about the minimum usable size for casual experimental work. Insert the copper-clad board in a thin, transparent bag, copper side up. For a small board, an ordinary plastic sandwich bag is fine. It is a good idea to tape the bag shut so the board can't slip.

In making our Kirlian "sandwich," the copper clad board is left in its bag and the film is placed outside the bag. The plastic of the bag provides some electrical insulation between the film and the copper-coated side of the board.

You will need fairly large raw sheets of film, not standard 35-mm film. The subject to be photographed is laid directly onto the negative, producing a life-size image. The film negative, therefore, must be large enough to accommodate the desired subject.

Make sure the wire connecting the copper-clad board to the V + output of the generator circuit is well-insulated. To be on the safe side, use the heaviest gauge wire you can find. Be very, very careful never to touch the copper-clad board when it is "live" (power connected). Make sure that the subject, film, and plastic sheets are all correctly oriented and positioned before applying any power to the circuit.

Because we are working with photographic film, you must take your Kirlian photographs in a dark or semidark room. Watch out for light leaks. You must work in the dark from the time you take the film out of its protective packaging until the photograph is developed. A low-light red bulb, as used in photographic darkrooms can certainly be a big help.

The high-frequency, high-voltage generator circuit (Fig. 7-1) will not work unless both switches S1 and S2 are closed. You can omit switch S1 if you prefer. It serves as a safety interlock. A normally open push-button switch should be used to supply power to the circuit. This makes it impossible to leave the high-voltage generator on when it is unattended.

When switches S1 and S2 are closed, power flows through transformer T1. A neon indicator lamp (I1) lights up, letting you know that power is flowing. To trigger the system and take the Kirlian photograph, while switch S2 is depressed, press switch S3. This causes the circuit to discharge, sending a high-voltage burst through the copper-clad board. Depress switch S3 only briefly. If you hold it down too long, you will overexpose your Kirlian photograph, and increase the risk of accidents due to the high power flow.

You will undoubtedly have to experiment by taking several test photographs until you find the best setting for potentiometer R2. This control adjusts the frequency of the discharge pulses. Remember, the exact frequency has an effect on the auras produced in Kirlian photographs. Bear in mind that the ideal frequency setting for different types of subjects might be quite different.

To take a Kirlian photograph with this system, set up the "sandwich" described earlier. Make sure the electrical connections are secure, well-insulated, and safe. Remember, never ground any living subject. Don't forget to turn out the room lights before removing the film from its packaging or you will ruin it. Turn on power interlock switch S1 (if used). Press and hold push-button switch S2. Make sure the indicator lamp (I1) lights up. While continuing to hold down switch S2, briefly press switch S3. Do not hold switch S3 down for long. Some experimentation might be required to determine the best exposure times for your particular setup. After releasing switch S3, release S2, and turn off interlock switch S1.

Without turning on the light, process the film, just like an ordinary photograph. If you do not have a photographic dark-room setup, place the film in a light-tight container before turning on the lights. You can then take the exposed film to any photographic lab and have them process it for you. You might want to tell them they are experimental photographs, so they won't assume the odd images are due to a light leak or some other defect.

Safety precautions

I can't overstress the fact that this is a high-voltage, high-current project, and it can be extremely dangerous if you do not take adequate precautions. Do not take foolish chances! Don't try to take shortcuts where safety is concerned. There is absolutely no reason for you to carelessly court the very real hazards of fire, or of a painful, potentially harmful, or even deadly electrical shock.

Make sure the entire circuit, and especially the ignition coil (T2), is well-insulated. When in doubt, use extra insulation, just in case. Use a heavy-gauge wire with thick insulation to carry the high-voltage output pulses to the copper-clad board. Do not touch the board or any other part of the system other than the operating switches when power is applied to the system. It is a good idea to build the project with a plastic (nonconductive) control panel. If a metal control panel is used, make sure it is completely electrically isolated from the circuitry, including ground. Do not use an earth ground for this circuit.

Grounding (circuit ground only) the subject improves the Kirlian photograph, but never, never ground a living subject. This almost guarantees a very painful shock. Serious injury or even death is highly likely. I repeat, DO NOT GROUND ANY LIVING SUBJECT!

Switch S1 is a safety interlock switch. It can be omitted, although its use is strongly recommended. If there is any chance that children might play with the equipment, an interlock switch is an extremely good idea.

Do not omit the fuse or attempt to bypass it. Do not use a fuse larger than 0.15 A. A 0.1-A fuse is recommended. Use fast-blow fuses only. Never use a slow-blow fuse in a high-power circuit like this.

When taking your Kirlian photographs, make sure no living creature is grounded; this includes yourself. Watch out for any possibility of accidental grounding. Wear rubber-soled shoes, if possible. Remember, you are working with high-voltage, high-current pulses. Take all precautions suitable to high-power work. Please don't take any chances. I guarantee it's not worth the risk.

❖ 8 Detecting UFOs

UNIDENTIFIED FLYING OBJECTS ARE OFTEN ASSOCIATED WITH THE New Age movement. Perhaps more than any other topic, there is considerable controversy about what UFOs are and what they mean, even within the ranks of New Agers.

On one level, UFOs do exist. They always have and always will. Whenever you see something in the sky and you don't know what it is, then, by definition, it is a UFO, at least from the observer's perspective. Someone else might know exactly what it is, so to that second observer, it is not a UFO.

Most people use the term "UFO" as a synonym for "flying saucer," and that is a completely different issue. While UFOs do exist, the existence of flying saucers remains highly questionable. And if flying saucers do exist, what are they?

Throughout the remainder of this chapter, I will use the term "UFO" in its popular, rather than its literal sense. When I speak of a UFO I am suggesting a craft of some sort or something else of unusual and inexplicable (by existing science) origin and nature.

Beliefs about UFOs

Belief in UFOs and even sightings and other related experiences are not restricted to New Agers. Such beliefs and experiences are certainly far from universal in the New Age movement. Some New Agers seem to have very little interest in the subject, while others are quite enthusiastic about it. There are many different beliefs about UFOs, which is not surprising because UFOs are, by definition, unknown items.

Probably the most common belief about UFOs is that they are spacecraft of some kind. Occasionally they are thought to be unmanned robotic craft, but usually they are assumed to be vessels carrying alien beings from some other planet or perhaps even another galaxy. A popular and commonly held belief is that the U.S. government knows the truth about UFOs and is covering it up, hiding the true evidence from the public. There are persistent rumors about the remains of a crashed UFO (and possibly one or more alien corpses) being hidden in some supersecret government installation. The most popular version of this story holds that the crashed UFO has been held by the government since the 1950s (when UFO sightings first became common).

Like most conspiracy theories, this one smacks of paranoia and doesn't really hold up to critical examination. It doesn't allow for human nature. Too many people would have to be involved—if only to guard, supply, and maintain the secret installation. If one of these people, perhaps after retiring, went public with the story, he'd certainly be an instant celebrity and make a lot of money on book and movie sales and TV appearances. The temptation to break silence would certainly be strong. Yet no one has come forth in all these years.

Could the government really be that powerful and all-controlling as to prevent any such leak? Recent history certainly shows that to be highly unlikely. Scandals like Watergate and the Iran-Contra affair have clearly demonstrated how easily a cover-up can be exposed. The exact detailed nature of what happened might be successfully covered up, but the fact that there is a cover-up and the general nature of what is being covered up is hard to keep secret.

I think it is more than reasonable to consider the "crashed flying saucer hidden by the government" story as a bit of folklore with little or no basis in reality. The story presupposes almost magical levels of control exerted by the government. It just isn't realistic at all.

But is it possible that UFOs are alien spacecraft from other planets? Well, sure. Anything is possible. But is it likely? Just what planets do these aliens come from? Surely not any of the planets in our solar system. Enough is known about them to virtually guarantee that no recognizable life-form (much less a race of intelligent, technologically advanced beings) could come from any planet in our solar system.

I suppose you can claim that they are totally different life-forms, perhaps not even carbon-based like earthly life. If such different life-forms do exist, it is highly unlikely that we'd recognize them as life or that they'd recognize us as such. Communication between such aliens and humans would almost certainly be impossible, or, at least, exceedingly difficult. Evolving so differently, the aliens would undoubtedly have very different technologies and goals. If they invented flying machines, the odds are enormous that the results would not be even roughly identifiable as such to earth beings. There is also, of course, the fact that multiple probes, and serious observation of the planets in our home solar system have not revealed any indication of life or activity.

If there are alien visitors to earth, they must come from some other solar system. Is this possible? Well, if their home planet surrounds a nearby star, it is theoretically possible, but they would have to be awfully patient beings and highly motivated to visit earth. Even the closest stars are several light years away. It would take several years to make the journey (one way), even if they traveled at the speed of light.

According to the theory of relativity (and some other scientific theories), it is theoretically impossible for anything to travel faster than the speed of light. The maximum speed for any practical vessel carrying any type of life, is certainly much slower than this. Again, it is possible that our existing scientific theories (including relativity) are completely wrong, but thus far, there is no good reason to assume this. New discoveries over the years have consistently supported these theories. Minor changes have been made, but the basic principles appear to be on very solid footing. We can conclude that faster than light travel (though it is convenient for science fiction authors) will always be impossible.

This reduces the chances of alien visitors from distant galaxies to such a small level that we can reasonably call it zero. Even from the closest stars the aliens would have to be in transit for several years each way. Why would they make such a long trip?

In the 1950s, much of the UFO folklore suggested that the aliens were of hostile intent, possibly scouting out earth for a future invasion, seeking materials, or slaves, or a new place for their growing population to colonize. Most modern UFO enthusiasts assume that the visiting aliens are not hostile. They might be neutral—perhaps they are performing a scientific study of some sort. In the New Age movement, they are usually (though not always) thought of as basically friendly. Their supposed intent is to help us in some way, guiding us to new discoveries (possibly of a mystic nature) or protecting us against ourselves, perhaps by helping to correct the environmental damage we have caused. Such helpful aliens certainly are an appealing concept, but they must be extremely dedicated helpers to travel all that way.

And their ways of going about things are certainly peculiar. If they want to help humanity, wouldn't it make sense for them to make themselves known to world leaders? Yet, if they wanted to keep their presence a secret, then why have they been sighted so frequently—almost always by accidental observers?

If we believe the pro-UFO claims, there are several different races involved. Sightings of "aliens" have often resulted in strikingly different descriptions. People who claim to have had direct contact with alien visitors all too often tell completely differing stories. What are the odds of such disparate races simultaneously making the long, hard trip to earth to play some bizarre variation of hide-and-seek?

It has been suggested that the visiting aliens come from a different dimension, rather than from another point in our space. This would solve the speed of light problem. Some theories in physics suggest multiple dimensions of some sort are possible, but any travel between them is based on principles completely unknown to modern science.

Some New Agers claim that UFOs are real and intelligently guided, but are not physical objects. Rather, they are points of contact with the spiritual plane. Sometimes UFOs are thought of as modern variations on ancient reports of angelic visitations.

It has also been suggested that the sighted UFOs are actually equipment being tested by the military. The trouble with this theory is that the "tests" have been going on for an awfully long time without leading to any known equipment. Even if it was a top-secret weapon, we run into the inherent problems of conspiracy theories, as discussed earlier. Too great a cover-up is required. There are too many possible sources for leaks, especially over such a period of time. No human agency could realistically be expected to maintain such universal silence over such long periods of time. Besides, the military admits to experimenting with other weapon systems and devices that have far more strategic importance than flying saucers. So why is the very existence of the government's flying saucers kept so secret? Even if the government could possibly cover up all traces of this activity, why would they bother?

Of course, skeptics can explain away all UFO sightings. Most (if not all) sightings, according to the skeptics, are misinterpretations of perfectly natural and fairly common things, such as weather balloons, swamp gas, cloud formations, and even airplanes. Such things certainly account for a large percentage of the UFO sightings that have been reported over the years.

Weather balloons are sent up by meteorologists to monitor atmospheric conditions. These balloons have highly reflective surfaces, can travel a considerable distance, and can catch light in unusual ways, allowing them to take on a wide variety of shapes. They sometimes hover in the air in what appears to be a very unnatural way. Sometimes it is not the weather balloon itself that is sighted, but a distant reflection of its shiny surface. This is likely to happen at night. The effect is very much like a searchlight shining against the dark sky. This permits the "UFO" to suddenly veer off in a different direction or suddenly disappear altogether. What has actually happened is that the weather balloon has gone into shadow or moved behind some obstacle, cutting off some or all of its reflection.

Swamp gas (primarily methane mixtures) can rise up and take on unusual shapes that can look like spacecraft or almost anything else. Moreover, swamp gas formations are often fluorescent, making the image appear to have "lights." Simple, everyday clouds have been mistaken for flying saucers too. A light source behind the cloud or reflecting against it can give the effect of a lighted craft

A UFO enthusiast showed me a book of photographic "proof" of UFOS. A few of the photos were a little puzzling, but the vast majority were clearly and unquestionably cloud formations. Many looked like illustrations from a meteorology textbook. One photo was unmistakably a small hurricane cloud forming over the ocean. In a few cases, I couldn't tell which of the several clouds in the picture was supposed to be the UFO.

For a long time I thought someone would have to be pretty stupid or gullible to mistake an airplane for a flying saucer, even at night. Then one night I got a firsthand demonstration of just how it can happen. A friend and I were sitting outside one night, talking and watching the airplanes come and go over the airport a couple miles away. We saw one right after it took off. The plane's outline was clearly visible at first, then all that could be seen were the lights on the plane. The plane banked, and because of our sight angles, it no longer looked like an airplane at all. Instead, it looked like a circular object with a ring of lights around its center. The illusion we saw is roughly illustrated in Fig. 8-1.

The "circular object" appeared to hover in the air for a few minutes, then (as the plane banked again) it appeared to zip off at an incredibly high speed and quickly vanished. The plane itself almost immediately reappeared, moving along a path different from the apparent direction of the illusionary "flying saucer." We saw the plane reappear only because we knew what we were looking at and we were watching for it.

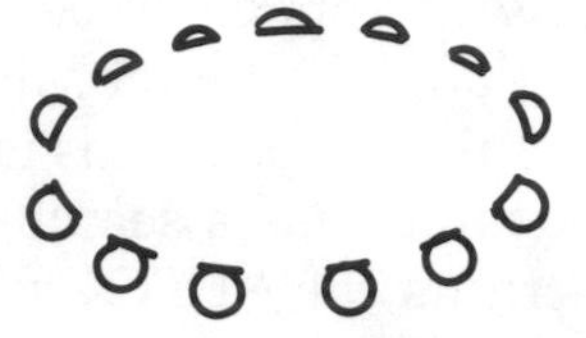

Fig. 8-1 At certain angles an airplane flying at night can look surprisingly like a flying saucer.

After this experience I can well understand how someone who happens to notice the lights in the sky after a plane has first banked could be convinced that he had sighted a spacecraft of some sort—a UFO. When the image vanishes, he looks where it appears to go and probably never sees the reappearance of the airplane as it continues on its way in a different direction. Our experience was unquestionably an optical illusion, but with different timing in our sighting, it could have been a very convincing UFO sighting.

I think that at least some UFO sightings are precisely this sort of optical illusion. But it would be naive to insist that such illusions account for all recorded sightings. Skeptics insist that all, or almost all, UFO sightings can be explained away along these general lines. Certainly, anyone but the most fanatic UFO believer will admit that a large portion (almost certainly a majority) of the reported UFO sightings over the years can be explained away in such fashion.

But there are a few sightings that don't fit any of these explanations. In fact, some sightings seem to be totally inexplicable from present knowledge. These are the important sightings and they can't be ignored, though the more rigid skeptics wish they could. It is quite likely that the unexplained sightings are actually manifestations of differing phenomena. That is, the explanation for mysterious sighting A might be totally different from the causes for mysterious sighting B.

Whenever there is something that is unexplained, it is an opening for scientific examination. Serious research has largely been abandoned, because worthwhile results have proved very hard to obtain. It is difficult to run scientifically controlled experiments on something when you don't know what it is, much less when or where it might show up.

Magnetic effects of UFOs

Many of the more mysterious and "unexplainable" UFO sightings are not just visual sightings; other phenomena are also involved. Frequently there are strong magnetic disturbances accompanying UFO sightings. These effects are quite inexplicable by ordinary means. Automotive speedometers, compasses, electric power meters, and the like have all shown odd effects during UFO sightings. They act like a very strong (sometimes fluctuating) magnetic field has been brought near them. Sometimes the visual sighting comes first and the magnetic phenomena is noticed later. In other cases, the equipment starts acting odd and then the UFO is visually sighted—occasionally by different people.

What causes these magnetic field effects? No one knows. They are rather rare, but they have been noted too often to be completely shrugged off. Is there some unknown natural phenomenon involved? Could whatever is causing the magnetic field effects also create visual effects or perhaps some sort of hallucinatory effect? Or is it possible that the UFOs really are mechanical devices or craft and the magnetic phenomona are side effects of their power sources (whatever they might be)?

Everyone has his own opinions and beliefs about such things. But hard, conclusive evidence is hard to come by. Certainly there has been nothing close to definitive proof, yet. It is possible that whatever is causing these effects is actually fairly common, but only the strongest, most unmistakable outbreaks have been noticed.

The next two projects, which make up the remainder of this chapter, are magnetic field detectors you can use to do your own experiments with such phenomena. Maybe you'll find a UFO of your own. These projects can also be adapted for more "down to earth" uses. They can be used to explore or monitor almost any type of magnetic phenomena, which, of course, are very common. A magnetic field surrounds every conductor carrying current, and the earth itself has its own fixed and varying magnetic fields, to name just two obvious examples.

Magnetic field detector project 1

The schematic diagram for our first magnetic field detector project is shown in Fig. 8-2. A suitable parts list for this project is given in Table 8-1. The secret of this project is its unique sensor, which is optoisolator IC1. There are special requirements for this component. You might have some difficulty finding exactly what you need, so I will try to give you a few different options you can try. An appropriate device for this purpose is the GE H1381 optoelectric coupler. Notice that it is an optocoupler, not a standard optoisolator. A similar unit can be substituted if this particular device is not available. The particular requirements are described below.

The unit is not completely enclosed as with most optoisolators. Instead, the body of the component has a central slot, giving it a sort of "U" shape. A small, thin object can be inserted into

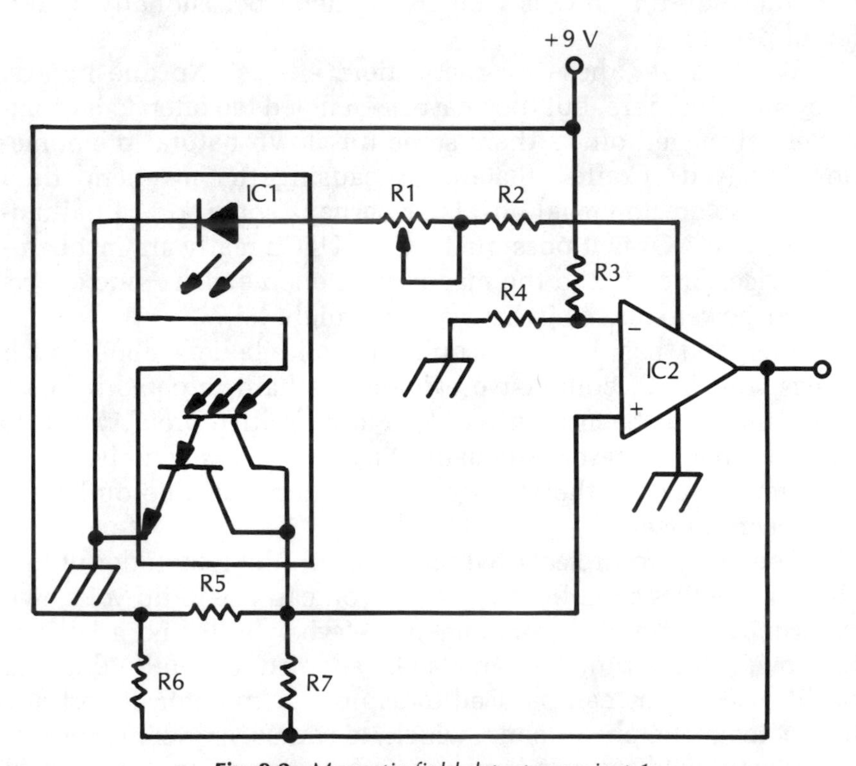

Fig. 8-2 Magnetic field detector project 1.

Table 8-1 Parts list for the magnetic field detector project of Fig. 8-2.

IC1	Optocoupler (npn Darlington phototransistor output) (GE H1381 or similar) (see text)
IC2	Op amp (741 or similar)
R1	5-kΩ potentiometer
R2	220-Ω, 0.25-W, 5% resistor
R3, R4	22-kΩ, 0.25-W, 5% resistor
R5	10-kΩ, 0.25-W, 5% resistor
R6	2.2-kΩ, 0.25-W, 5% resistor
R7	100-kΩ, 0.25-W, 5% resistor

the slot, blocking the flow of light from the internal LED to the internal phototransistor, as illustrated in Fig. 8-3. You can call it a mechanically interruptable optoisolator.

The output device for the optocoupler is an npn Darlington phototransistor pair. If you must use a unit with just a single phototransistor output (it must be an npn type, of course), the second

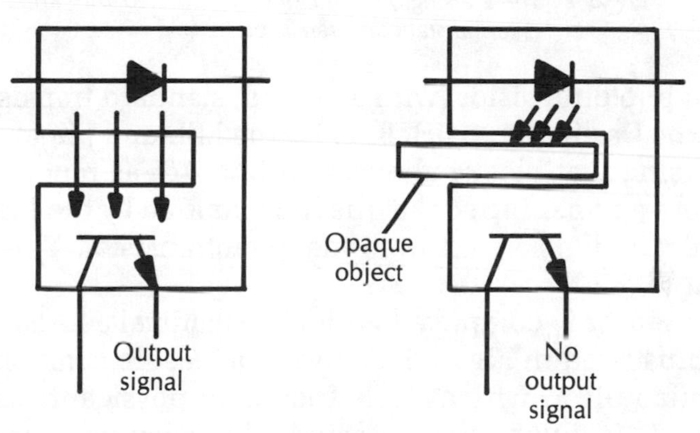

Fig. 8-3 A small, thin object can block the flow of light through an optocoupler.

transistor of the Darlington pair can be added outside the optocoupler, as illustrated in Fig. 8-4. For best results, the external transistor (which is not a phototransistor) should be reasonably well-matched to the optoisolator's internal phototransistor. With the proper selection of transistors, there is no electrical difference between this hybrid version and the single unit indicated in Fig. 8-2.

Optocouplers are not as easy to find as optoisolators. If worse comes to worse, you can build one of your own from a separate

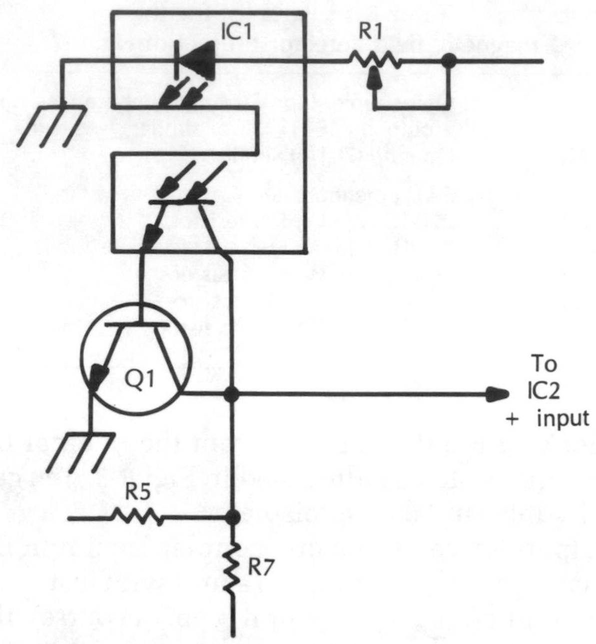

Fig. 8-4 The Darlington pair can be completed with an external standard transistor.

LED and phototransistor (with a second, standard transistor completing the Darlington pair). Enclose the LED and phototransistor in a housing that blocks all external light (or as much as possible). Leave an opening so that the light emitted by the LED can be mechanically blocked off from the phototransistor. This is illustrated in Fig. 8-5.

The sensor is completed with a mechanical compass assembly, as illustrated in Fig. 8-6. A magnetic needle is mounted on a free-turning pivot, with as little friction or physical resistance as possible. This much is just a standard mechanical compass. In the presence of an external magnetic field, the magnetic needle will physically align itself to the detected magnetic field.

Strips of paper, or some other opaque, lightweight material are attached to either end of the compass' needle. Be careful not to add too much weight or the compass assembly will not function properly. It won't be able to turn with sufficient freedom.

The optocoupler sensor is mounted so that when the compass needle is in certain positions, one of the paper wings moves through the optocoupler's slot, blocking off the internal light connection, as shown in Fig. 8-6B. In other compass positions, the light flow is unimpeded, as shown in Fig. 8-6A.

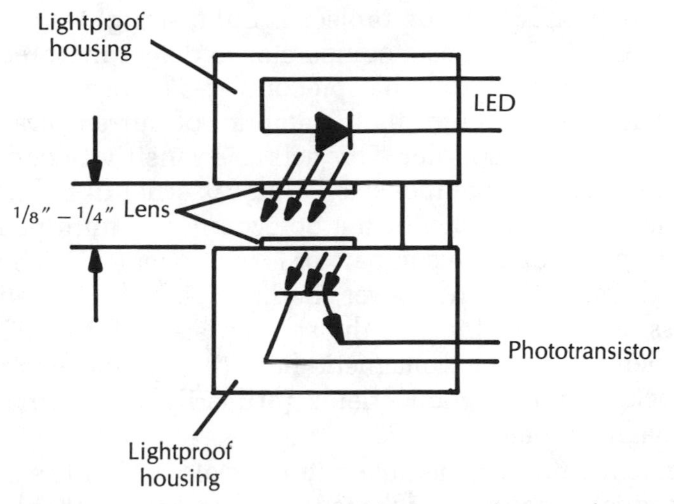

Fig. 8-5 *In a pinch, you can build your own homemade opto-coupler assembly.*

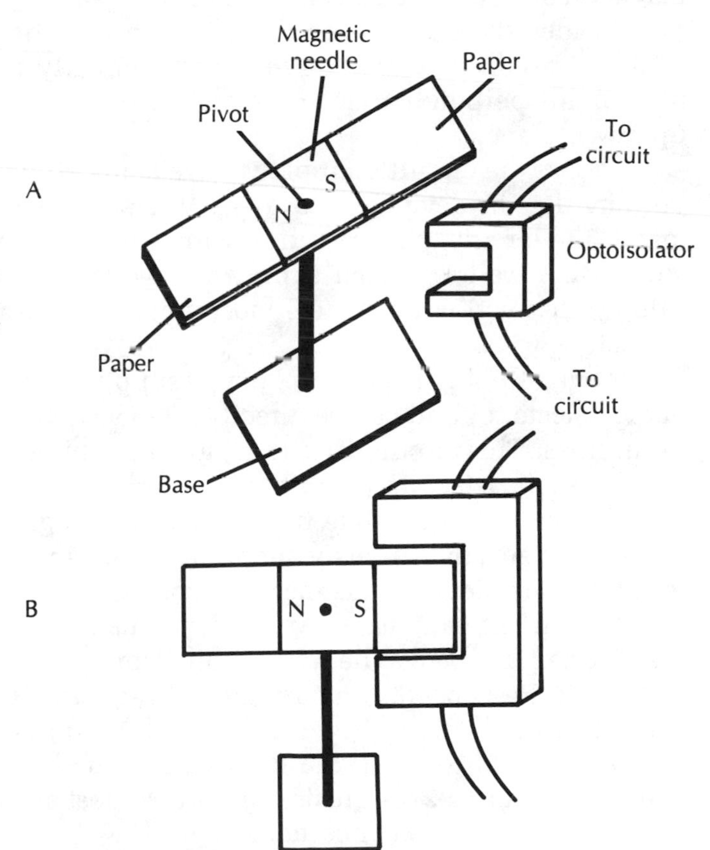

Fig. 8-6 *The mechanical compass assembly for the magnetic field detector project of Fig. 8-2.*

The remainder of this project is quite straightforward. The op amp (IC2) is wired as a comparator to detect the presence or absence of the signal from the optocoupler. The resistance setting of potentiometer R1 controls the amount of current available to the LED in the optocoupler. This acts as a sensitivity control and can be used to help compensate for any external light leakage.

When the light path in the optocoupler is unimpeded, the output from the op amp comparator is low (near 0 V). The output goes high (near V+) whenever the light path is blocked by the compass needle moving into the appropriate position. This happens when the detected magnetic field (if any) exceeds the normal, background magnetic fields (primarily the earth's north-south magnetic field).

There are no intermediate output levels. Either the output is low (magnetic field not detected) or it is high (magnetic field detected). The output signal can be used by many analog and digital circuits, depending on your specific application. This magnetic detector circuit can drive CMOS gates.

Besides the obvious north-east-south-west orientation, the compass needle can also be mounted vertically to detect magnetic fields perpendicular to the earth's surface—that is, in the up-down plane.

Choose the circuit's supply voltage to match the desired output circuitry or devices. While +9 V is suggested in the schematic, the circuit will work fine with a wide variety of supply voltages. A well-regulated supply voltage (of whatever specific value you select) will offer the most reliable performance for the project.

If a tightly regulated +5-V power supply is used, the output of the detector circuit can be directly interfaced to most computer input lines. This circuit drives a single input bit to the computer. Several such detectors (at differing locations or angles) can be used in parallel as a single input byte. The programming of the computer can continuously monitor, track, and keep detailed records of any detected magnetic anamolies.

The circuit can also drive a relay. This is illustrated in Fig. 8-7. Diode D1 protects the relay's coil from back-emf, which can damage it. Resistor R8 is a current-limiting resistor and its value should be small. It might not be needed in all cases. The series dc resistance of R8 plus the relay coil itself should be at least 300 Ω. If you use a 220-Ω resistor for R8, almost any relay will do, as far as resistance is concerned.

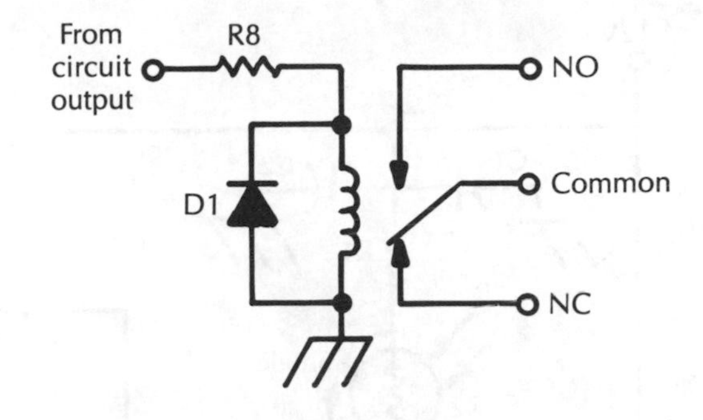

Fig. 8-7 The magnetic field detector project of Fig. 8-2 can be used to drive a relay.

This magnetic detector circuit cannot drive a large relay directly. The comparator's safe output is limited to about 20 mA (0.02 A). If you need to drive a larger load, you can use a small relay to drive a larger relay, or you can add a simple transistor current-amplifier stage between the comparator's output and the relay coil.

Magnetic field detector project 2

Our next project is another magnetic field detector, but it is designed to operate according to totally different principles. The schematic diagram for this project appears in Fig. 8-8 and a suitable parts list is given in Table 8-2. This project is a somewhat simplified variation on a design used in some professional magnetometers.

The sensor in this project is a photoresistor (R2) and a CRT (cathode-ray tube). CRTs are used in oscilloscopes and as picture tubes in television sets. For this application, a fairly small screen will do just fine and will probably be more convenient, reducing the weight, size, and general bulk of the completed project.

You can buy a brand new CRT, but it is cheaper to ''cannibalize'' one from an old, discarded oscilloscope or TV set. Often you can buy an old TV set quite cheaply at a yard sale, especially black-and-white TVs. Either a color or a black-and-white tube will work fine in this project. The color tube offers no particular advantage in this case.

If given a choice, a CRT designed for use in an oscilloscope is probably the best choice. They are smaller, lighter, and require less power, and often have a calibrated grid printed directly on the screen face. But if you want to cut corners and save a little money, you might do better with a used picture tube from a small-screen television set.

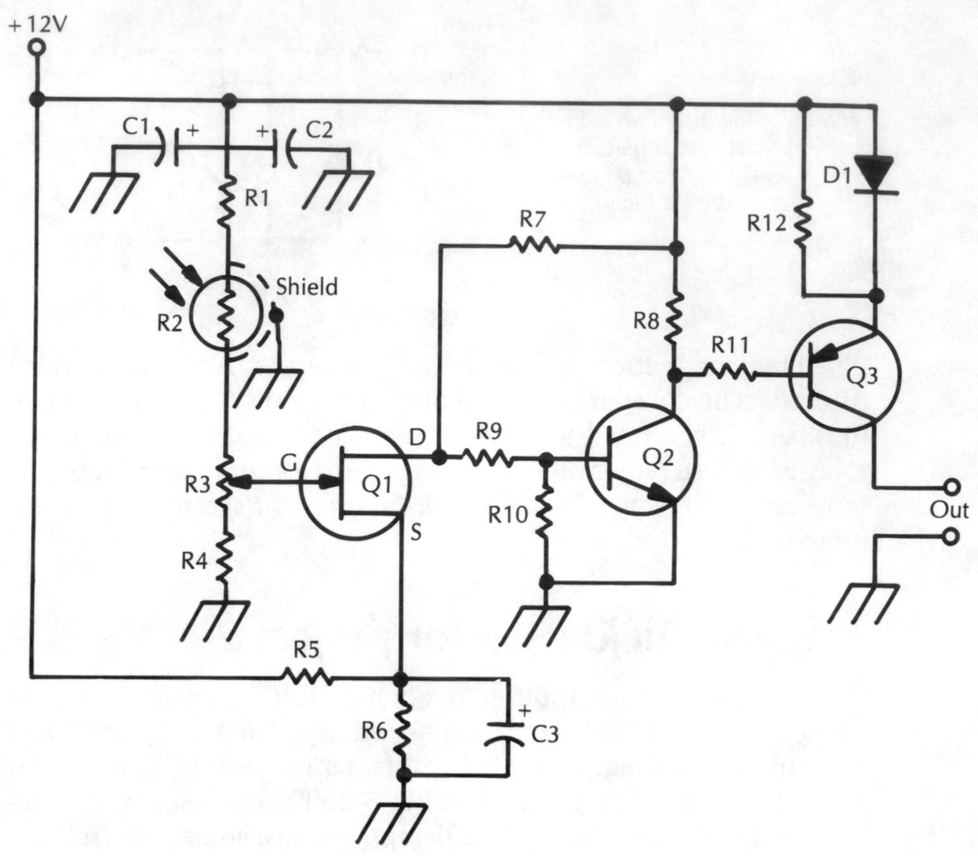

Fig. 8-8 Magnetic field detector project 2.

Table 8-2 Parts list for the magnetic field detector project of Fig. 8-8.

Q1	N-channel FET (Radio Shack RS2035, HEP-F0010, or similar)
Q2	npn transistor (Radio Shack RS2009, HEP-S0011, or similar)
Q3	pnp transistor (HEP-S0012 or similar)
D1	Diode (Radio Shack RS1104, HEP-R0050, or similar)
C1, C2	470-μF, 50-V electrolytic capacitor
C3	47-μF, 50-V electrolytic capacitor
R1	100-kΩ, 0.25-W, 5% resistor
R2	Photoresistor
R3	500-kΩ potentiometer
R4	220-kΩ, 0.25-W, 5% resistor
R5, R10	1-kΩ, 0.25-W, 5% resistor
R6, R11	330-Ω, 0.25-W, 5% resistor
R7, R9	6.8-kΩ, 0.25-W, 5% resistor
R8	4.7-kΩ, 0.25-W, 5% resistor
R12	12-Ω, 0.5-W, 5% resistor

You can even get away with using a partially defective CRT, such as one that pulls in the pictures along the edges or has a blurred picture. If you find a used CRT with a bad yoke, but it is otherwise OK, that is perfect. The yoke assembly is not used in this project.

The photoresistor is affixed to the center of the CRT's screen, as illustrated in Fig. 8-9. If you are using a defective CRT with an off-center image, that is no problem. You are concerned only with the image center, not the geometrical center of the screen itself. The photoresistor should be positioned so that the undeflected (no input signal) electron beam is focused directly on it.

In most normal applications, the electron beam is moved across the screen to display an image. This is done via the yoke assembly. The yoke is a set of electromagnets around the neck of the CRT. Electrical signals applied to the yoke coils create magnetic fields that bend or deflect the electron beam so it strikes a

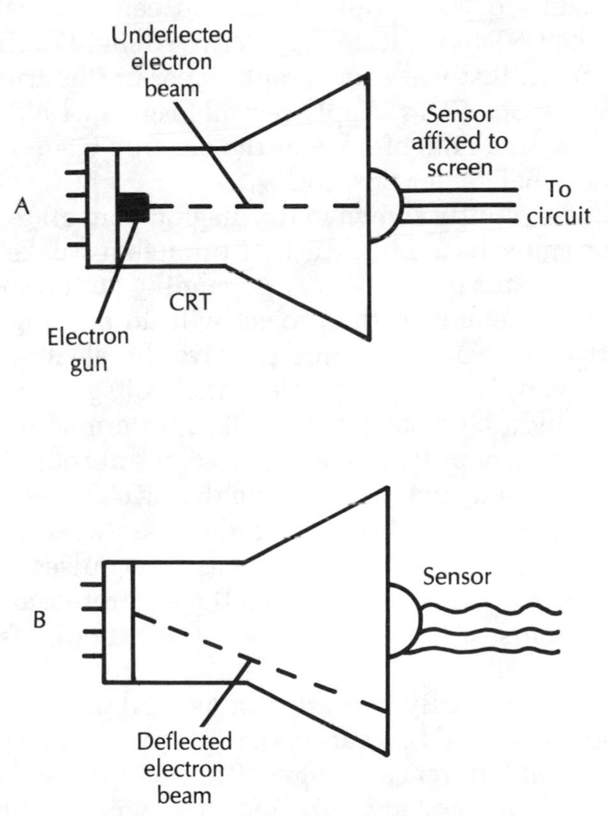

Fig. 8-9 *The photoresistor is fastened directly to the center of the CRT's screen in the magnetic field detector project of Fig. 8-8.*

different spot on the screen. In this project, the yoke assembly is removed and discarded. If a magnetic field crosses the CRT, it will deflect the electron beam accordingly. The photoresistor no longer ''sees'' the glowing spot at the center position, and the detection circuit is triggered. This sensor is very sensitive, even to relatively small magnetic fields.

Potentiometer R3 is used to calibrate the instrument so it triggers the circuit correctly. Adjust this control so the circuit is not triggered when the electron beam is not deflected (spot centered). The ordinary, constant magnetic field of the earth (and any other consistent magnetic fields in the environment) are automatically compensated for when you determine where the electron beam strikes in the absence of any additional angle. Use a small permanent magnet held near the neck of the CRT to manually deflect the electron beam away from the photoresistor. Adjust potentiometer R3 to trigger the circuit under these conditions.

If suitable to your application, you can also calibrate the screen of the CRT to indicate the relative strength of the detected magnetic field. Just use some graph paper or the grid found on many oscilloscope CRTs. Apply several magnetic fields of known force, and make a note of how far the electron beam is deflected from its nominal center position.

It is not explicitly shown in the diagram, but all relevant supply voltages must be fed into the CRT through its socket pins. The exact requirements here will vary depending on the specific CRT you use. Just remember, the project will do nothing without a sufficiently high voltage source to drive the electron gun. You also need to apply an appropriate signal voltage to the electron gun. This might be a steady, dc voltage, or a medium- to high-frequency string of pulses. In either case, it's purpose is to keep a continuous lighted spot displayed on the CRT's screen. A pulsed signal source will probably cause a little less wear and tear and potential overheating of the electron gun. The pulses should have a fairly high frequency so that when the electron beam is turned off between pulses, the detector circuit will not be falsely triggered.

If you are primarily interested in using this project to detect UFOs, you might find that various earthly magnetic fields constitute a serious interference problem. The solution is to shield the CRT from these unwanted magnetic fields, while leaving it sensitive to the desired magnetic disturbances you want to monitor.

Because UFOs will presumably be overhead, and most interfering magnetic fields will probably originate near the ground, one solution is to place the CRT within a steel box or other enclosure with the top open. In this way, the detector will only "see" magnetic field disturbances from the sky.

Of course, because high voltages are required to operate the CRT, adequate shielding and insulation is an absolute must. Don't take foolish chances. This is particularly vital if the sensor assembly is to be mounted outdoors. Protect all electronic components from the elements. If mounted outside, it is a good idea to use plastic or some other magnetically permeable material to close off the open end (top) of the CRT's steel enclosure. This will keep out rain and debris. It will also keep out curious fingers that could receive a severe electrical shock.

The output of this detector circuit can drive CMOS digital gates or analog circuitry. A computer interface of some sort can probably be built fairly easily. The specifics depend on your particular computer system and your intended application. An A/D (analog-to-digital) converter might be necessary.

The output of the detector circuit can drive a relay, as in the preceding project. Again, depending on your specific intended application, a latching relay might be desirable. Once the circuit is triggered by detection of a magnetic field, the output remains activated until the latching relay is manually reset, usually with a simple push button. This circuit can also drive a small meter, an electronic buzzer, or some other alarm signaler.

9 ❖ Crystals

WHILE THESE LAST TWO CHAPTERS DON'T FEATURE ANY MORE actual projects or circuits, I think you will find the subject matter interesting. I will be making some final comparisons between New Age and science or technology (particularly electronics). Probably the most common and familiar object in New Age circles is the crystal, which is not fundamentally dissimilar from the crystals often used in electronic circuitry.

What is a crystal?

To understand exactly what a crystal is, we first need to take a brief look at atomic theory. At one time atoms were thought to be the smallest possible particles. Splitting an atom was unthinkable, because there could never be anything smaller than an atom. In fact, the very word "atom" means "indivisible." As science and technology improved, it was discovered that atoms really are divisible after all. They can be split into smaller components. An atom is actually the smallest particle that is recognizable as a specific element. If you split, say, a gold atom, you will no longer have gold, but a collection of nonelemental particles.

The chief components of an atom are the proton, the neutron, and the electron. Every electron is just like every other electron. There is no difference between a gold electron and a lead electron. The same thing is also true of protons and neutrons. What distinguishes one element from another is the number of each of these components the atom contains. (We now know that even these subatomic particles can be further split into still smaller particles.)

Normally, each atom has the same number of electrons as it has protons. Because protons have a positive electrical charge and electrons have an equal negative electrical charge, this equality of numbers means that the normal atom, as a whole, is electrically neutral. Under some circumstances, an atom might lose an electron, making it a positive ion. Under other conditions, it might pick up an extra electron, making it a negative ion. It is also possible, and very common, for an atom to share one or more of its electrons with one or more other atoms. As we shall see, this is very significant for the formation of crystals.

The number of protons in an atom is called the atomic number. (Normally the number of electrons also equals the atomic number.) The atomic number defines what the element is; for example, hydrogen has an atomic number of 1 (one proton and one electron) and oxygen has an atomic number of 8 (eight protons and eight electrons). The protons and the neutrons are clumped together into a central mass called the nucleus. The electrons orbit around the nucleus, somewhat like planets orbit around the sun, as illustrated in Fig. 9-1. This image isn't strictly accurate, but it will do for our purposes.

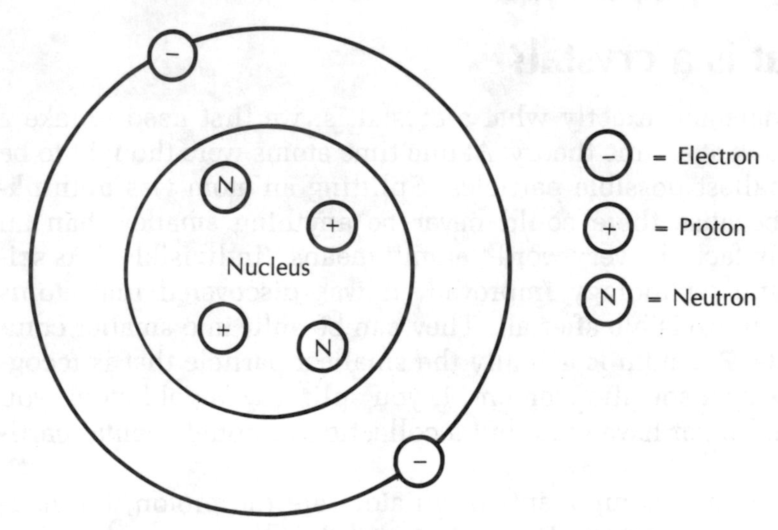

Fig. 9-1 *In an atom, the electrons orbit around the nucleus.*

Notice that the electrons cannot orbit at any random distance from the nucleus. They always form definite rings. Moreover, these rings have a very definite pattern in the maximum number of electrons each ring can contain. The first ring, the one closest to the nucleus, can hold one, or, at most, two electrons. If the atom has three electrons, two will be in this innermost ring, and the third will be in the second ring a little farther out from the nucleus. Each ring must be completely filled to its maximum before a new ring is started. Of course, the outermost ring in any atom might have less than its maximum number if there aren't enough electrons to fill it. The second ring can hold up to 8 electrons. The third ring can hold up to 18; the fourth, 32; the fifth 50, and the sixth, 72 electrons. No known atom has more than 6 electron rings. This information about electron rings is summarized in Table 9-1.

Table 9-1
Number of electrons that can be held in each electron ring.

Ring number	Maximum number of electrons
1	2
2	8
3	18
4	32
5	50
6	72

Now let's imagine a typical atom, which doesn't have its outermost ring completely filled. It essentially has one or more "holes" or "spaces" for extra electrons. Nature abhors a vacuum, so the atom tries to fill up these spaces if it can. If it is close to another atom with a comparably incomplete outer ring, it will try to "steal" these outermost electrons from the second atom. Meanwhile, the second atom is trying to steal the first atom's outermost electrons. In effect, the two (or more) atoms share their outermost rings, forming a very tight bond. The atoms have grouped together into a molecule. If the atoms within a molecule are not of the same element, the resulting substance is a compound that has characteristics unlike any of its component elements.

This can be made clearer with a specific example. An ordinary hydrogen atom has one electron, in the first ring. This outermost (and only) ring has "space" for one more electron. An ordinary oxygen atom has eight electrons. Two fill up the first, innermost ring, leaving six in the second ring. The second ring can hold up to eight electrons, so it has "spaces" for two more electrons. Two hydrogen atoms will easily and readily join with a single oxygen atom to fill up their outermost rings. Each hydrogen atom shares one electron with the oxygen atom, so it "thinks" it has two electrons in its outermost (only) ring. The oxygen atom, at the same time, is sharing the two electrons from the two hydrogen atoms, so it "thinks" its outermost (second) ring is full with eight electrons. These three atoms combine in a very neat compound molecule (H_2O). The bond is so tight, no other atom (of any element) can force its way into the molecule. In effect, the three atoms in the molecule are sharing a sort of double figure-eight outermost ring made up of eight electrons, as illustrated in Fig. 9-2. This particular combination, of course, is water. Notice that water is very different from either hydrogen or oxygen separately.

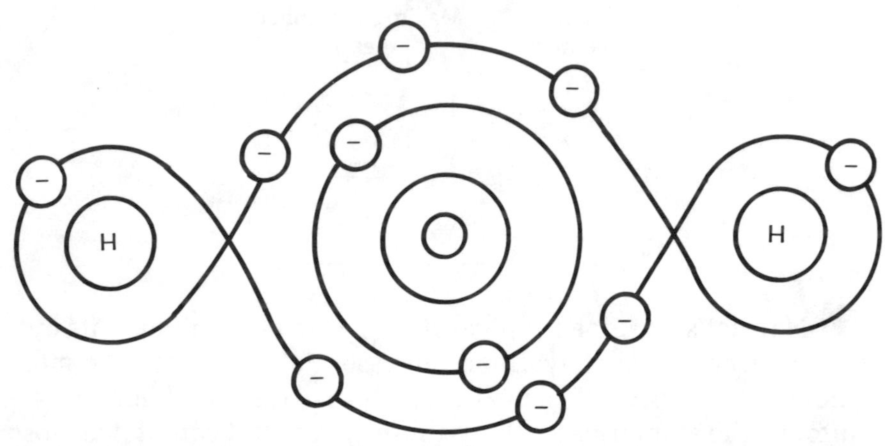

Fig. 9-2 *In a water (H_2O) molecule, two hydrogen atoms and an oxygen atom share their outermost electrons.*

Atoms with unfilled outermost electron rings join to form molecules whenever possible. But not every combination is possible or equally likely. Atoms "like" to get together only in combinations that result in a filled outer ring. Same element atoms can combine to form elemental molecules. For example, two hydrogen atoms combine to form a hydrogen molecule.

Some elements have a full outermost ring. For example, helium has an atomic number of 2, so it has two electrons, which completes the first ring with no electrons left over. An atom like this won't join with any other atom to form molecules. It is complete unto itself. Such "uncooperative" elements are sometimes called the "noble gases" because they occur naturally in the form of gases. In addition to helium, the noble gases are neon, argon, krypton, radon, and xenon. A neon molecule consists of just a single neon atom and nothing else. Neon does not appear in any other compounds.

Certain substances have a special, regular pattern in the way their molecules are put together. This special pattern is called a *crystal lattice*. Practically all inorganic (nonliving) solids occur in crystalline form. Even materials such as iron, copper, and aluminum are crystalline in nature. A piece of iron is made of a great number of molecular crystals lying in random positions throughout the material. A substance composed of a large number of molecular crystals is called a *polycrystalline* material.

Certain substances are less random. The molecular crystals are arranged in regular, rather than random positions. Such a substance is a true crystal. Quartz is one of the most familiar examples. In fact, often when the word "crystal" is used, it is referring to a quartz crystal. New Age references to crystals almost always refer exclusively to quartz crystals.

Another very common type of crystal is common table salt. If you examine table salt under a strong magnifying glass or microscope, you will see that the small grains appear as tiny cubes. Compare this with black pepper. The grains of pepper are irregularly shaped and dissimilar from one another, unlike the striking uniformity exhibited by the salt. The salt is crystalline, but the pepper is not.

Salt molecules are made up of sodium atoms and chlorine atoms in a crystalline lattice. The arrangement of atoms in a salt (sodium chloride) crystal is shown in Fig. 9-3. The lines between the sodium and chlorine atoms represent the shared electrons or chemical bonds that hold the crystal together. Due to the way in which these bonds form, every pure crystal will be like every other pure crystal. This precise, repeating arrangement of atoms within a crystal is called the crystal lattice. The physical properties of a material (hardness, tensile strength, etc.) are determined to a large degree by the lattice structure of the material.

Quartz is one of the most common elements on earth and it is very crystalline in nature. Unlike many other types of crystals, quartz crystals often have a visible crystalline structure, even at the macro level. You can see the crystal structure with your naked eyes, without the need for special equipment like a magnifying glass or a microscope. Usually, unless some other type of crystal is specifically mentioned or implied by the context, it is reasonably safe to assume that when the word "crystal" is used, a

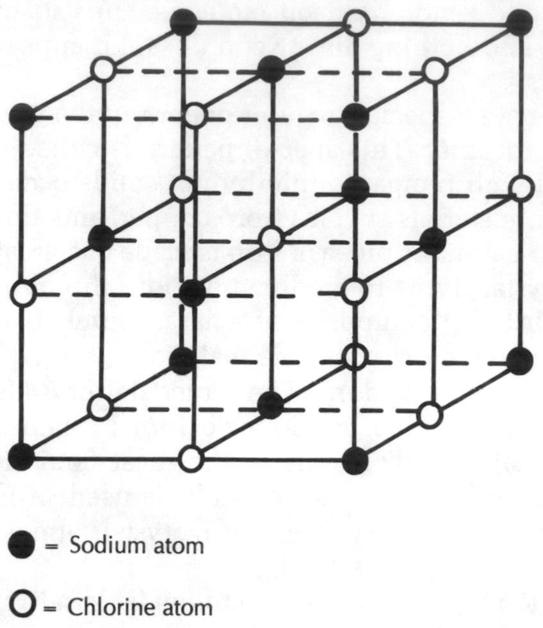

● = Sodium atom

○ = Chlorine atom

Fig. 9-3 Common table salt exhibits a typical crystalline structure.

quartz crystal is meant. New Age crystals are almost always quartz.

Crystals in electronics

Quartz crystals are also widely used in electronics. As an amusing sidebar, do you remember when digital watches first began to catch on? The most expensive models were advertised as being "crystal controlled" or "quartz controlled." Eventually, as prices began to drop, the phrase "quartz controlled" seemed to be limited to top-of-the-line models, as if it was a special feature. All digital watches are crystal controlled, and it is always a quartz crystal that is used. The phrase still apparently holds some marketing appeal, even though it is no longer reserved to expensive watches. I just bought a digital watch for 99 cents and the packaging proudly announced it was "quartz controlled."

At any rate, in an electronic circuit, only a thin sliver of a quartz crystal is used. Figure 9-4 shows the basic internal structure of the electronic component known as a "crystal." A thin, precisely cut slice of quartz is sandwiched snugly between two metal plates. These plates are held in tight contact with the crystal with small springs. This entire assembly is enclosed in a metallic (usually) case that is hermetically sealed to keep out

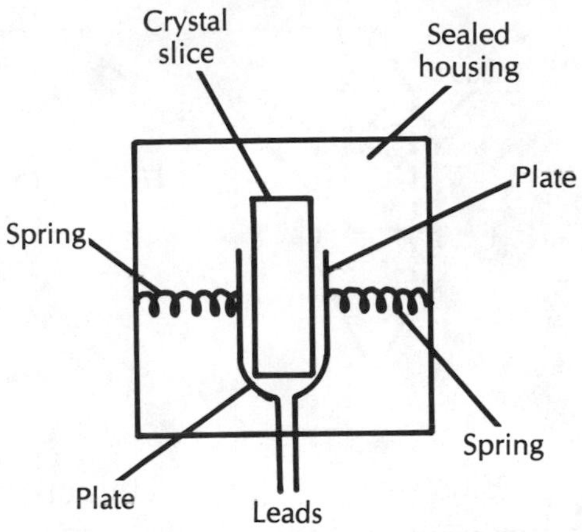

Fig. 9-4 *The internal structure of a typical crystal used in electronic circuits.*

moisture and dust, which can damage the delicate crystal slice. Electrical leads connected to each of the metal plates are brought out from the case for connection to an external circuit. The schematic symbol for such a crystal is shown in Fig. 9-5.

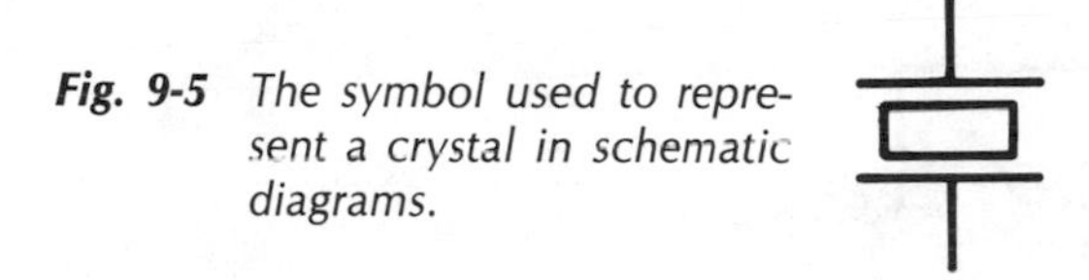

Fig. 9-5 *The symbol used to represent a crystal in schematic diagrams.*

A crystal works in an electronic circuit because of a phenomenon called the *piezoelectric effect*. Two axes pass through a crystal. One is called the X axis, and it passes through the corners of the crystal. The other is called the Y axis, and it is perpendicular to the X axis, but in the same plane. Figure 9-6 shows these two axes in a typical crystal, looking down through the top of the crystal.

Thanks to the piezoelectric effect, if a mechanical stress is placed across the Y axis of a crystal, an electrical voltage is produced along the corresponding X axis. Similarly, if an electrical voltage is applied across the X axis, a mechanical stress is created along the corresponding Y axis.

One important result of this piezoelectric effect is that a crystal can easily be made to ring (vibrate) or resonate at a specific frequency, under certain conditions. In practice, this is easiest to do with a thin slice or sheet of quartz crystal.

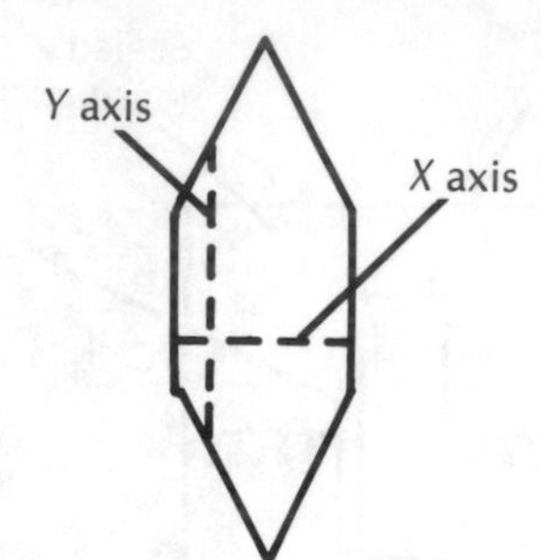

Fig. 9-6 The piezoelectric effect depends on the X and Y axes in a crystal.

Electrically, a crystal functions as if it was a network of capacitances, inductances, and resistances. The equivalent circuit for a typical crystal is illustrated in Fig. 9-7. Depending on how the particular crystal was manufactured, it can operate either as a series resonant LC network (Fig. 9-8) or a parallel resonant LC network (Fig. 9-9). In a series resonant LC network, the impedance (ac resistance) is at its minimum value at the resonant frequency. A parallel resonant LC network works in exactly the

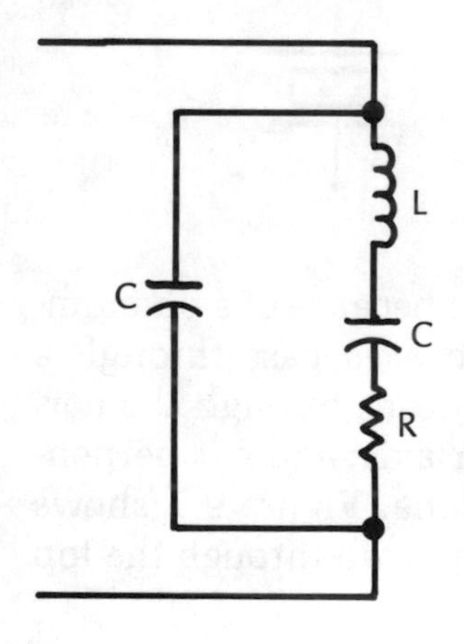

Fig. 9-7 The equivalent circuit for a typical crystal.

Fig. 9-8 Some electronic crystals are designed to take the place of a series resonant LC circuit.

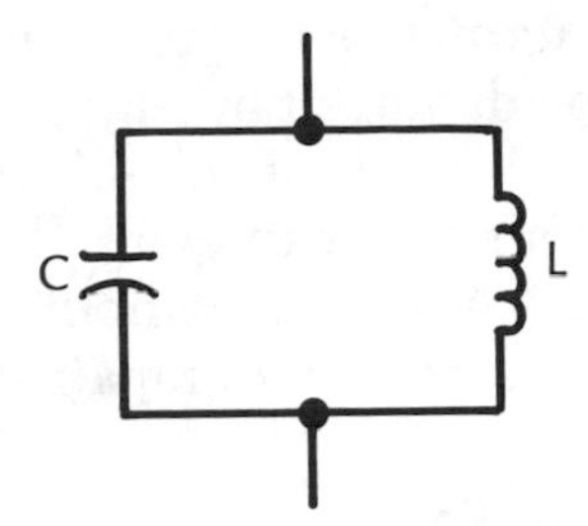

Fig. 9-9 *Some electronic crystals are designed to take the place of a parallel resonant LC circuit.*

opposite manner. The impedance is at its maximum value at the resonant frequency. Generally speaking, a crystal designed for series resonant use cannot be used in a parallel resonant circuit, or vice versa.

The resonant frequency of a crystal is fixed and is determined by the thickness and size of the crystal slice. The resonant frequency of a crystal is in the RF (radio frequency) range. The resonant frequency will be at least a few dozen kilohertz (1 kHz = 1 000 Hz). Some crystals have resonant frequencies of several megahertz (1 MHz = 1 000 000 Hz).

Crystals can be made to resonate at integer multiples of their main resonant frequency. These multiples are called *overtones*, or *harmonics*. The primary frequency is known as the *fundamental*. For example, a crystal designed to resonate at 150 kHz (150 000 Hz) (the fundamental) will also resonate (but to an increasingly lesser degree) at 300 kHz (second harmonic), 450 kHz (third harmonic), 600 kHz (fourth harmonic), and so forth. Notice that there is no ''first harmonic,'' because one times the fundamental is equal to the fundamental. The resonance effect becomes steadily less pronounced at higher harmonics.

Crystals are usually more expensive than separate capacitors and coils, but the resonant frequency of a crystal is much more precise and stable. LC resonant circuits often drift off-frequency (that is, the components change their values somewhat), particularly under changing temperature or other environmental conditions.

While not as temperature sensitive as discrete capacitors and inductors, crystals can change their resonant frequency slightly in response to changing temperatures. Therefore, when very high accuracy is required (as in broadcasting applications), the circuit is enclosed in a special crystal oven, which is nothing more than a temperature-controlled enclosure.

Another important advantage of crystals is their inherent reliability. They have a much lower failure rate than comparable circuits using discrete capacitors and coils. However, a crystal can be damaged by high overvoltages or extremely high temperatures. A severe mechanical shock (such as being dropped onto a hard surface) can crack the delicate crystal element. If this happens, the entire component must be discarded and replaced. There is no way to repair it.

New Age crystals

Now, just what does all of this have to do with New Age? Crystals are one of the most common elements in New Age beliefs and practices. The crystal can almost be considered a symbol for the entire New Age movement. New Age beliefs and claims about crystals range from the plausible to fantastically far-fetched fantasies.

Some say crystals are alive or even intelligent. It has been claimed that crystals can aid or cause healing (both physical and psychological), or that they can capture and focus thoughts, desires, and emotions. Some New Agers use crystals to purify water and improve the taste of wine. Crystals can supposedly increase or amplify ESP powers, set up or strengthen psychic shields for protection against negative influences, permit communication with the dead or other disembodied spirits, ease astral projections, and energize people or even machines. It has been claimed that crystals can perform such useful wonders as improving gas mileage, controlling the earth's magnetic fields, and even serving as powerful energy sources. According to many versions of the legends, the mythical ancient civilizations of Atlantis and Lemuria supposedly used crystal power as their primary energy supply.

This is quite an extensive and incredible set of claims, yet it scarcely scratches the surface of New Age beliefs about crystals. It is doubtful that anyone really believes all of these things about crystals, but each of these beliefs (and many others) is commonly held throughout New Age circles.

Much has been written and published about crystals and "crystal power." Often much is made of historical connections to ancient uses of crystals by various once mighty civilizations. Putting aside mythic "lost" civilizations such as Atlantis and Lemuria, Egypt is generally considered as probably the most important and generally influential of these crystal-using ancient civilizations. It has been claimed by some authors that the pyramids were once capped or perhaps faced with crystals. This is certainly possible, but there doesn't seem to be any particular evidence to support such claims beyond the beliefs of these authors and their followers.

Much of what is claimed in New Age writings about the ancient use of crystals is historically questionable. Many authors claim very precise knowledge about ancient rituals and other matters for which absolutely no documentation exists. Just how do these authors come by their information? Certainly crystals were widely used by ancient civilizations, especially in religious, magical, and ritual contexts. This is not surprising, because crystals are naturally beautiful, and their inherent regularity of form and structure would undoubtedly appear mysterious and magical to a prescientific culture. Crystals look almost like artifacts or manufactured rocks, so it is only natural to expect prescientific peoples to assume that there is something somehow special about these decidedly unusual stones.

But is it science?

Some New Age authors have accepted the widespread use of crystals in modern electronics to be evidence of their claims about crystal power. The immense potential power of crystals is just fractionally used. But technology and modern science are on the path of rediscovering what the ancients once knew about crystals and their powers.

Unfortunately, it seems as if virtually all of the New Age authors writing about crystals don't understand (or prefer to ignore) what modern science does know about these stones. Pseudoscientific theories about crystal power are all too often based on distortions of current scientific knowledge and end up as nonsensical goobledygook that sounds very impressive to the uninformed. This is regrettable, because there is always a possibility that there might be some glimmer (or more) of truth in some of the New Age claims about crystals. But many good hypotheses are buried amidst the often incoherent and ridiculous superstitions.

In the next few pages I will compare popular New Age beliefs and current science on crystals, concentrating on four basic areas: vibrations, crystal energy, crystals and healing, and living crystals. I will make every effort to be as unbiased as possible in these discussions, pointing out possible avenues of profitable scientific research to explore the more promising New Age hypotheses. You'll notice that I try to ignore the more spiritual aspects of New Age crystal lore. This is because such things are, by definition, outside the realm of testable science and technology.

Crystals and vibrations

As mentioned in chapter 1, the notion of vibrations is an important component of New Age philosophy. Vibrations are held to be of particular importance in the case of crystals. Their regular, patterned internal structure makes crystals exceptionally sensitive to vibrations at specific frequencies.

Unquestionably, crystals are prone to vibrations at specific frequencies. That is precisely why they are used in oscillator and tuned circuits. But New Agers carry the idea somewhat farther. They are not really interested in the electrical vibrations of an oscillator circuit, but in psychic or mental vibrations. Supposedly a crystal can pick up vibrations from thought patterns. These vibrations can then be stored within the crystal or, in some cases, amplified by the crystal. These thought vibrations can be transmitted to or from the crystal over distances of up to several feet away.

But just what are these thought vibrations that are said to be interacting with the crystal? Apparently, they are not directly related to anything that has yet been measured in any laboratory. It has been suggested that these New Age thought wave vibrations are related to the brain waves that have been detected and measured by science. This sounds good at first glance, but this "explanation" runs into severe problems when we look at the frequencies involved.

The thin slices of quartz crystal used in electronic circuits resonate (vibrate) at a specific frequency defined by their physical characteristics—primarily, the shape and size of the crystal. Usually the resonant frequency is several thousand or several million cycles per second. In New Age use, the exact size and shape of the crystal doesn't seem to be selected for the desired "thought frequency." Usually, other factors such as color are considered more important by New Agers. Many New Agers seem to believe that almost any crystal can be used for any thought frequency, depending on how it is "programmed." A crystal is programmed by projecting appropriate thought waves at it.

Brain wave frequencies are much, much lower than the resonant frequencies of crystals in electronics. Measured brain waves are normally less than a couple dozen cycles per second. There is no known way to get a crystal to electrically vibrate at such a low frequency. Then there is the added problem that brain wave signals have very low amplitudes. They are barely measurable or even detectable at the surface of the skull. They can easily be drowned out by background interference.

It is possible that there is more to brain waves than what science has discovered so far. The currently known brain waves are an electrical phenomenon, and they are certainly less than fully understood. Is it possible that the brain waves discovered thus far are just side effects of some more powerful but scientifically unknown phenomenon?

Perhaps someday science will discover that there are other sorts of vibrations involved in thought, and it is not completely inconceivable that crystals might react to and interact with such vibrations. Such a discovery would also lend strong support to other New Age ideas, such as telepathy and clairvoyance (see chapter 4). However, there appears to be no scientifically valid evidence in support of such a hypothesis at this time.

From a scientific viewpoint, at present there is no reason to believe that crystals interact with "thought vibrations." Of course, there have been very few, if any, valid scientific studies to test such ideas. Many such experiments are certainly possible. Have one person "program" a crystal with a particular thought or emotion, then, have another person (isolated from the first) try to pick up the "stored vibrations" from the crystal. A strong double-blind procedure is necessary to make such a study reasonably valid. It shouldn't be possible for the person conducting the test to inadvertently tip off the "receiving" subject through unspoken cues. For example, someone might handle a crystal that has been "programmed" with something negative differently than one "programmed" more positively. Any researcher in direct contact with the "receiving" subject should not know how the specific crystal used in the test has been "programmed."

A few control tests with "unprogrammed" crystals should also be included in the study to help identify subjects who are just "lucky" guessers.

Crystal energy

Many New Agers believe that crystals are powerful energy sources. They believe crystals have an energy field extending several feet around them. This is one reason why so many New

Agers wear crystal jewelry or keep crystals placed at strategic locations around their home and office. They believe the crystal's energy helps revitalize them and offers some sort of psychic protection. One popular ''proof'' of crystal energy is that when sensitive people hold a crystal in their hand and concentrate, they can feel warmth or a tingling sensation emanating from the stone.

If crystals do emit any sort of energy, it is apparently of an entirely unknown type. Modern technology never uses crystals as energy sources. Crystals do, however, act as energy transformers. They can convert mechanical energy into electrical energy, and vice versa. Notice that nothing is created in the crystal, only the type of energy is altered.

What we are talking about here, of course, is the piezoelectric effect, which was discussed earlier in this chapter. Briefly, the piezoelectric effect means that if a mechanical stress is placed across a *Y* axis of a crystal, an electrical voltage is produced along the corresponding *X* axis. Similarly, if an electrical voltage is applied across an *X* axis, a mechanical stress will be created along the corresponding *Y* axis. No energy is created or emitted in the piezoelectric effect. The effect takes place entirely within the crystal. None of the energy ''leaks'' out of the crystal. No measurable or detectable energy field of any known type surrounds a crystal.

What about the sensations some people claim to feel while holding crystals? Are they lying? Probably not. But the mind is a funny thing. They might be feeling what they expect to feel. One potential experiment is to see if they experience the same effects if they hold a very convincing imitation crystal, which they think is the real thing. In other words, this would be a variation of the placebo test. Many New Age phenomena might have strong placebo effects involved. But no one can know for sure which do and which don't until serious scientific experiments with strict controls and double-blind precautions are performed, preferably by multiple research groups.

It is possible that the ''energy'' felt by some people when holding a crystal is related to some biofeedback phenomenon. Biofeedback was discussed in some detail in chapter 3. Some people can adjust their surface blood flow rate, which would be experienced as localized heat or a tingling sensation.

Crystals and healing

Alternate approaches to healing and health are a major part of the New Age movement. It seems only natural that this would be blended together with the use of crystals. It is an obvious combination of two of the most popular New Age themes.

Many New Agers believe crystals are powerful healing tools. There are several variant ideas of how and why they supposedly aid healing. Some hold that it has something to do with the mysterious vibrations that crystals are assumed to emit or resonate with. Others say the crystals absorb or draw away the negative disease energy. Still others claim that only a properly programmed crystal will work. Positive thoughts of good health are stored in the crystal for transmission to the patient. Ideas along these lines can get quite sophisticated and complicated. Different programming (or perhaps different types of stones) are used for different diseases. The healing crystals (and sometimes other gem stones) are also alleged to aid in resistance against disease or other problems (physical, psychological, or even spiritual negative forces are said to be held off, reduced, or modified by the appropriate stone or stones). This is why so many New Agers wear crystal jewelry and certain other gems. In New Age circles, jewelry is rarely worn simply for decorative purposes. Certain types of jewelry might be flatly rejected as having some sort of negative effect.

New Age authors often make much of the use of crystals by shamans in primitive cultures. Certainly crystals have been used this way in many cultures. The visually intriguing crystalline structure in itself is enough to lead a superstitious mind to suspect it must be magical. Shamanistic use of crystals was certainly far less widespread and important than many New Age authors imply or claim. In some cases, to be sure, crystals were among the shaman's chief tools, but as often as not, if crystals were used in healing ceremonies, they served a relatively minor role. Certainly many crystal artifacts have been found. New Age crystal enthusiasts seem to ignore the fact that artifacts made of crystals or other stones are much more likely to have survived to be found by archeologists than equally (or even more) important artifacts made of wood or animal skins or feathers.

The argument to accept the healing qualities of crystals because such properties were known to ''ancient science'' seems highly inconsistent when the same people don't accept everything else ancient shamans used. Why are crystals accepted as "scientifically" valid, while other parts of the ancient healing ceremonies (if known) are simply ignored? The most common and usually the most important part of shamanistic healing ceremonies was making a lot of noise to frighten away the evil spirits that were thought to cause illness.

While scientific studies into the alleged healing effects of crystals, with proper controls and double-blind procedures, have been rare, it seems quite likely that any observed effect is probably due to the placebo effect. The crystals work because the people involved believe they will work. The placebo effect has been widely studied, though it still hasn't been fully explained. If patients are given an inert substitute (such as a sugar pill) and are told it is a powerful drug, 30% to 50% (in most studies) respond to the placebo as if it really was a drug. The placebo itself is selected so that it can't have any effect of its own; the patients's response is attributed to his own beliefs and expectations.

It seems likely that there is a strong element of the placebo effect in many New Age ideas. I believe it will work, so my belief makes it appear to work. It is important to remember that the effects achieved by the placebo are real, there is only misdirection as to the cause. Internal effects (beliefs and expectations) are doing the work attributed to an outside agent (the placebo).

Anecdotes of specific cases where crystal healing (or anything similar) worked as predicted are therefore rather weak as scientific evidence. There is no way to tell from the evidence how much (if any) placebo effect is involved. Because there is no scientifically valid reason to attribute the healing results to the crystals, it seems reasonable to assume that the crystals are most likely functioning as placebos.

However, there is some evidence that psychic healing might not be complete quackery in all cases. Sometimes the laying on of hands or other psychic treatments do appear to work. In many cases, the patient is not faking or showing hysterical, psychosomatic symptoms. Often someone who has not responded to traditional medical treatment does improve or is even entirely cured after treatments by a psychic healer.

While the scientific evidence in support of such ideas is extremely weak (thus far, anyway), there is some possibility that some sort of ESP effect might exist that relates to healing. In this case, perhaps the use of crystals serves a symbolic function. It is difficult for most people to deal with things on a purely psychic or mental level. Crystals, as physical objects, give them something to emotionally hang on to.

The regular structure and many facets of a crystal make it a good meditation aid. The subject can focus on the crystal to aid in concentration to maximize his mental powers (whatever they might be). Perhaps a paper clip will work just as well, but a crystal is prettier and more symbolic.

Are crystals alive?

Perhaps the most striking and far-fetched of all New Age ideas is the claim that crystals are alive. On a direct, literal level, this assertion is utterly ridiculous. Crystals most certainly do not qualify as biological life. However, remember in chapter 1, I mentioned that New Age stresses the ultimate interconnectedness of all things in existence. In a sense, the universe is alive, so all of its component parts must also be alive. Of course, this includes crystals as much as anything else. But crystals seem to hold a special status here. Some (not all) New Agers insist that crystals are uniquely living entities of some sort. Some have gone so far as to suggest that crystals are actually the original source of all life.

To some extent, this belief appears to be based at least partially on the fact that crystals, unlike most other rocks, can grow. Of course, a crystal's supposed vibrational characteristics and its link with human thought patterns also plays an important part in this curious belief in living crystals. But the resemblances of crystals to biological life are very slight and superficial at best. Biological life forms take in nourishment and convert it into energy and body parts for growth. What is taken in and cannot be used is excreted as waste. A crystal can absorb certain chemicals that interact to increase the size of the crystal. This is growth of a sort, but it isn't really very similar to biological growth. And the only way a crystal can ''reproduce'' is by being broken into pieces. Of course, this has nothing at all to do with biological reproduction.

In the ordinary, biological sense, crystals are most definitely not alive. Of course, the definition of ''life'' can be loosened enough to apply to anything at all, but this is not the way most people use or understand the word. In all probability, the ''living crystal'' idea originated as a metaphor for metaphysical concepts, but it has come to be taken literally.

Psionic generators

Originally, I thought I might be able to include an exciting project in this chapter. In some New Age literature, I had read of something called a psionic generator or sometimes simply a ''black box.'' This device is most often mentioned in connection with crystals, which seemed to be an important component in the design. A psionic generator is a machine with electronic parts that is often interfaced with a computer and is used to amplify psychic energy—that is, the thoughts and emotions of the user.

While I had some doubts that such a device would do all it was claimed to do, I felt it would be an interesting project to experiment with. Unfortunately, as I continued my research, I wasn't able to find any coherent detailed description of a psionic generator, much less anything resembling useful plans for constructing such a device. Instead I found a lot of pseudotechnical mumbo jumbo apparently written by people who didn't have the slightest idea of what they were talking about. A typical set of ''construction plans'' stated that the number and placement of the potentiometers and capacitors is decided by personal preference. In other words, there is no actual circuit. The electronic components are put together for aesthetic appeal, not for any particular electrical connections. Somehow these randomly arranged components magically get together and amplify signals no one has ever detected. From the available evidence I've come across in my research, I can't help but conclude that psionic generators are complete nonsense, of use only to con artists and stage magicians. Such things give the entire New Age movement a bad name.

❖ 10 The quantum connection

UNDOUBTEDLY MANY (MAYBE EVEN MOST) READERS OF THIS BOOK still feel that many of the New Age concepts discussed here are hopelessly far-fetched and ridiculously unscientific. Certainly it is true that hard, scientific evidence to back these concepts is, to date, scarce or even nonexistent. But in many cases, there are intriguing bits of "soft" evidence—not enough to prove anything, but enough to raise questions. It goes against the basic idea of science to flatly ignore all such questions. It is probable that scientific investigation into many of these New Age concepts won't turn up any useful results. Many New Age beliefs are likely to be disproved, but isn't it possible that at least some aspects of the New Age might hold at least a few kernels of truth, as yet undiscovered in the laboratory? Certainly the scientists of today don't know everything, and the honest ones among them will freely admit that.

In the earlier days of this century, it might have seemed much easier to make blanket statements about what is and what isn't real. But quantum physics has thrown so much of what we thought we knew into a cocked hat, that it is foolish and completely unscientific to prematurely declare any idea false without thorough investigation.

Quantum physics was mentioned briefly in chapter 1. In this final chapter, I will take a closer look at the peculiar mysteries and paradoxes of quantum mechanics and the questions and possibilities raised by these startling discoveries. It is important to realize that as bizarre as much of it sounds, quantum physics does make sense and is consistent, but not at all in the same way as classical physics.

I can barely scratch the surface of quantum physics here because it is a very complex subject. If you want to learn more, many nontechnical books on quantum physics have been published in the last few years. If some of it doesn't seem to make sense, don't worry too much about it. It's a little like a jigsaw puzzle. You can't even guess what the complete picture will look like until you've managed to fit quite a few pieces into place. At some point, you'll find a key piece and suddenly things will start to make a lot more sense. And remember, our knowledge of quantum mechanics is still terribly incomplete. Nobody (including award-winning scientists) understands it completely. Some aspects of quantum physics might even be beyond human comprehension because it is so unlike everything we are used to in the everyday world.

New and old physics

In one sense, quantum physics has completely disrupted the basic assumptions of traditional physics. On the other hand, the laws and principles of classical physics are still true as ever, on the level they were created in. It is vital to realize that the peculiar quantum effects I will be discussing here have been observed only on a very tiny, submicroscopic scale. They occur only with the incredibly tiny particles (or pseudoparticles) that make up the electrons, protons, and neutrons which make up the atoms which make up everything we see. On the ordinary, day-to-day level we are used to, quantum effects have not been observed, and the predictions of classical physics hold true. So classical physics is still quite valid, and is far from being thrown out lock, stock, and barrel.

But more levels have been added. On the subatomic scale, physical principles don't work the same way, only smaller. Instead, totally different principles apply on the subatomic scale. Often, discovering what is happening on the subatomic, quantum scale significantly changes our perspective of what is really happening on the classical macro scale. The events observed remain the same, of course, but the explanation of why things happen a certain way are colored by quantum physics. The classical concept of cause and effect has been greatly complicated. Purely classical cause and effect explanations still give valid predictions, but such explanations are ultimately less precise and detailed than the full quantum explanations. It's a little like rounding off the results of a complex mathematical equation. For most practical purposes, 3.14 will give perfectly adequate results for the value of pi, even though that is not the true value, which extends to an infinite number of decimal places. But how could anyone solve any equation using the full value of pi? It has to be rounded off for practical applications.

So, in a real sense, the laws of classical physics are useful approximations for practical use. They work, but they don't tell the whole story. As it turns out, many of the ''hidden'' aspects of the whole story, as revealed by quantum physics are incredibly surprising and often more than a little paradoxical.

The peculiar nature of quantum particles

One of the hardest things for most scientists (and even nonscientists) to accept is that quantum physics has conclusively demonstrated that purely objective observation is impossible. Much of classical physics is based on the idea of an objective observer monitoring what is going on without having any effect on the process. Good physics experiments are carefully designed to prevent any unintended effects from the observation process.

On the quantum level, it has been clearly demonstrated that the effects of observing a phenomenon cannot be eliminated. The very act of observing something directly and unavoidably affects what is being observed. An example of this was briefly touched on in chapter 1. You can measure a quantum particle's velocity or its position, but not both. It's not just a matter of being unable to perform both observations simultaneously; you can't even measure one and then the other. Once the velocity of a quantum particle has been fully determined, that particle no longer has any definite position. Somehow the very nature of the particle itself has been altered by the act of observation.

It's not that some particles have a definite velocity and others have a definite position. Before any measurement is made, the particle has the potential for both. Whichever measurement is made first, will give the desired results. But then the second measurement (of the other factor) won't work.

But it is possible to compromise. You can know a little about a particle's velocity and a little about its position, but the more exactly you know its position, the less you can determine about its velocity, and vice versa.

All available indications are that this is not just an illusion or a technological limitation of modern measurement techniques, but it is actually a necessary element of the nature of the quantum particles themselves. Are many New Age ideas really weirder or more absurd than this?

To make things even stranger and more paradoxical, quantum particles can't even seem to "decide" if they are matter (particles) or energy (waves). This curious dual nature was first discovered with photons, which are units of light. The same thing has been found to apply to all particles on the quantum level.

In some ways a photon looks and acts like a particle—a little physical piece of something, an object. For example, if it strikes something, it might knock off an electron or two. This is how the photoelectric effect (used in solar cells, photoresistors, and many other light-sensitive electronic components) works. On the other hand, photons also behave as waves of energy. Measured in certain ways, it appears that light is made up of tiny particles of matter. Measured in other ways, it appears that light is pure energy, made up of waves, not matter.

Which is a photon really? Particle of matter or wave of energy? Both, and neither. On the quantum level, the matter/energy distinction breaks down completely. This was presaged by Einstein's famous equation, $E = MC^2$ where E is energy, M is mass (measurement of matter), and C is the speed of light. This equation tells us that matter can be converted into energy, or vice versa. Quantum physics tells us matter and energy are essentially just different phases of the same thing.

All quantum particles exhibit this curious duality, and all of the component parts of every atom are made up of these indeterminate quantum particles. And, of course, all matter is made up of large collections of atoms. So is there ultimately any such thing as matter at all or is it just "disguised" energy? Many New Age philosophies claim that material existence is basically an illusion, and everything is made up of energy or vibrations. Are the New Agers saying the same thing as the quantum physicists, albeit in different terms?

The double-slit experiment

A classic quantum experiment demonstrates this peculiar dual nature of quantum particles. I will describe this experiment as a "thought experiment," as if it could be done on the everyday macro level. This is simply to avoid dealing with a lot of technical details that aren't really essential for our purposes. But be aware that actual quantum-level versions of this experiment have been performed and have repeatedly given the mind-boggling results I describe here.

The experiment requires a source of quantum particles—just for convenience, I will assume they are photons. Some distance away, there is a screen. The photons are aimed directly at this screen. An image appears on the screen, indicating where photons have hit it. As described so far, we have the setup shown in Fig. 10-1. The photons spread out and cover the entire screen.

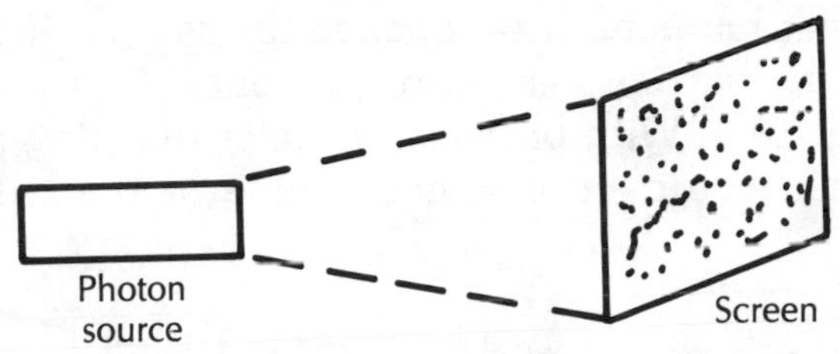

Fig. 10-1 *Photons fired at a screen will tend to spread out and evenly cover the screen.*

For the next stage of the experiment, a barrier is added between the photon source and the screen. This is not a complete barrier, but a card with a single slit cut in it, as illustrated in Fig. 10-2. The results are pretty much the same as before, except the slit focuses the photons. On the screen, there is a dark bar forming the image of the slit, with only a few stray "hits" outside this area, as illustrated in Fig. 10-3.

So far we have nothing special here. The photons are acting just as if they were ordinary particles. If we increase the size of

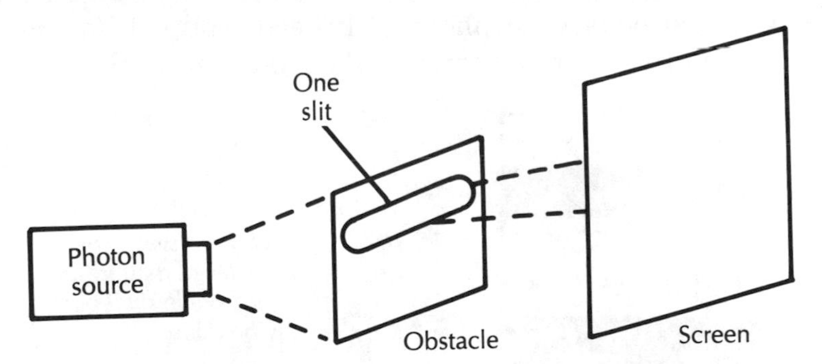

Fig. 10-2 *In the first step of the experiment, you pass the photons through a single slit.*

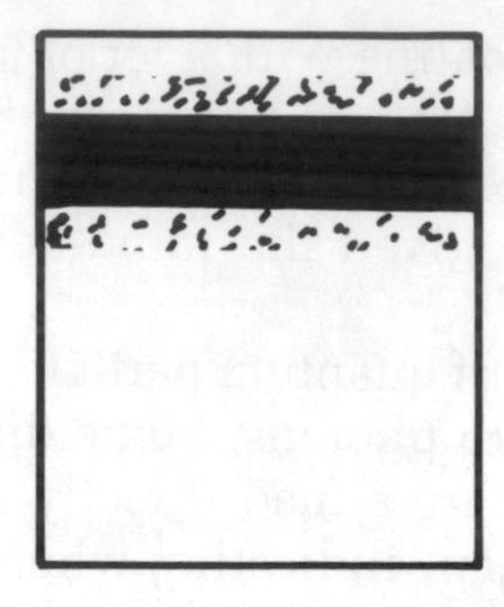

Fig. 10-3 The single slit limits most of the photons to just a portion of the screen.

the experimental elements, we would get pretty much the same result with ping pong balls instead of photons.

Now, what happens if we change the setup a little and put two slits in the barrier instead of just one? This new setup is shown in Fig. 10-4. With physical particles like ping-pong balls, nothing much would be changed. Some of the balls will go

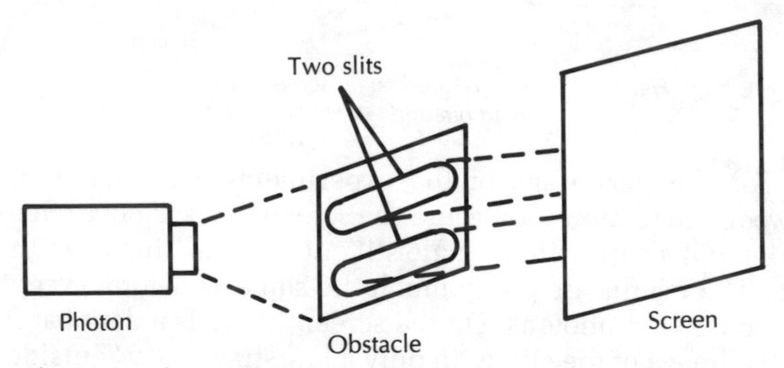

Fig. 10-4 When a second slit is added to the experiment, the results start to get weird.

through the upper slit and some will go through the lower slit, creating two stripes on the screen, as shown in Fig. 10-5. But this isn't what happens with photons. Instead, you get an interference pattern, like the one shown in Fig. 10-6. Fewer photons than

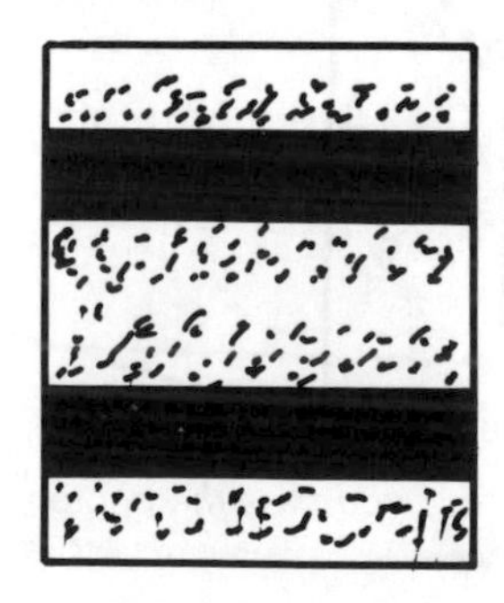

Fig. 10-5 If the photons were physical particles of matter, the effect of a single slit would simply be doubled.

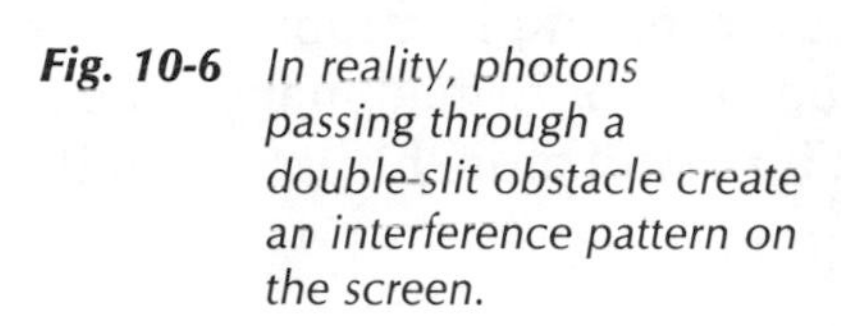

Fig. 10-6 *In reality, photons passing through a double-slit obstacle create an interference pattern on the screen.*

expected hit the screen at certain spots. There are places on the screen that are somehow ''avoided'' by the photons. Block off one of the slits and you get more ''hits.''

If the photons are physical particles like ping-pong balls, then these results make no sense. Consider the experiment from the point of view of any single photon. It must go through either the upper slit or the lower slit. After all, how could any single particle manage to go through both slits? So how in the world does the photon ''know'' both slits are open so it can avoid the dark portions of the interference pattern on the screen?

An obvious solution to this paradox is that some of the particles bump into each other and deflect each other from the dark spots. Sounds logical, but it doesn't hold up experimentally. If you slow down the photon stream, so that only one photon at a time reaches the screen, you get the same interference pattern, it just takes longer for it to build up over time. This is impossible for any physical particle.

So, apparently, the photons are not really object-like particles at all. Then they must be waves, right? Waves can pass through both slits at once and can interact to create interference patterns. Does that clear up the paradox? Not at all. If you were dealing with a wave, it would hit the entire screen at once, not just a single point. But each of our photons (or other quantum particles) does hit only one specific point on the screen. If you fire just one photon through the experimental setup, there will be just one tiny lighted spot on the entire screen, as if the photon was an object or particle.

So, this experiment tells us that a photon (or other quantum particle) is not a true particle (object) and it is not a wave (energy). It is sort of both, but neither. With one slit open, the photons act like particles, but if two slits are open, they act like waves. The nature of the photons is dependent on how the experiment is set up—that is, how you observe them.

Take some time to think over the implications of this experiment, which has been repeated and confirmed many times. Once you think about it for awhile, it is even more bizarre than it seems at first glance.

Quantum movement

On the ordinary macro scale, objects move smoothly, crossing through the space between their beginning and ending points. For example, let's imagine an object moving along the line shown in Fig. 10-7, from point A to point F. The object obviously has to move through all of the intermediate points; that is, after leaving point A, it must pass through points B, C, D, and E before it reaches point F.

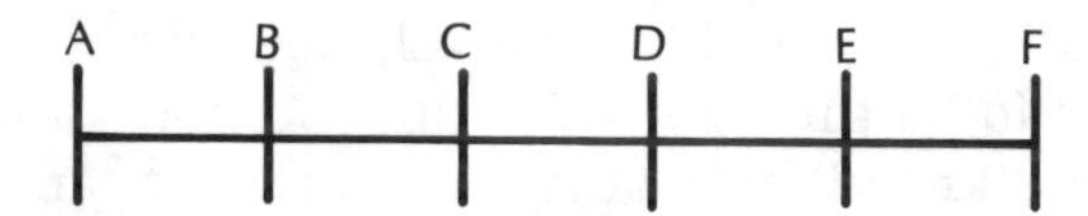

Fig. 10-7 *Quantum movement is not linear. Quantum particles actually leap directly from position to position, without crossing the intermediate space.*

Now, all of that might well seem far too obvious to be worth mentioning. But it isn't the way things work on the quantum level. Quantum particles do not pass through all of the intermediate space. Instead, they can instantly jump from point to point, without ever crossing the space in between. For example, a quantum particle might start at point A on our line, then jump to point C, and then jump again to point F. At no time does this particle occupy points B, D, or E. It just skips over them altogether. This effect is often called a *quantum leap*.

An example is the distance of an electron orbit from its nucleus. It can't be at just any distance. It must be at one of several discrete and specific distances, just like a digital signal must be either high or low, and never anything in between. A simplified drawing of an atom is shown in Fig. 10-8. Normally the electron is in orbit A. If the atom is excited by an external energy source, it jumps to orbit B. The electron can never be between orbit A and orbit B. What's more, when the atom is excited, the

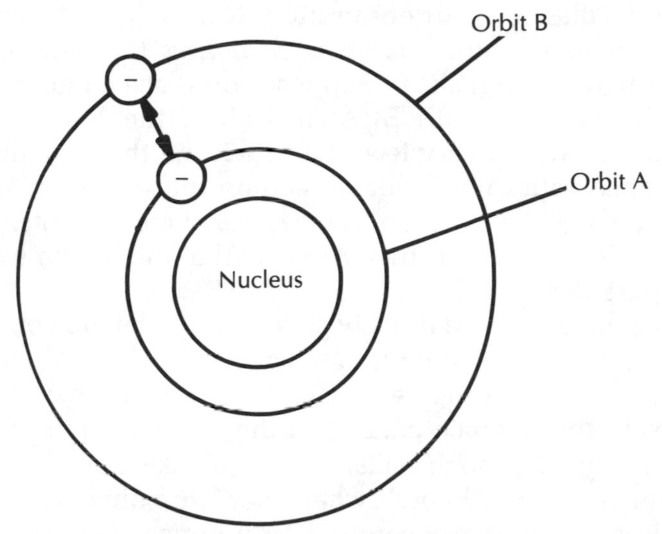

Fig. 10-8 *Each electron must be either in orbit A or in orbit B, but never in between the two orbits.*

electron instantly jumps from orbit A to orbit B. It never crosses through the space in between.

The indeterminacy principle

One of the most confusing and most frequently misunderstood aspects of quantum physics is the *indeterminacy principle*. I have already touched on this principle somewhat in the problems of determining either a quantum particle's velocity or its position. Quantum particles follow statistical laws. It is impossible to predict what a specific particle will do or when it will do it, even though the possibilities are usually severely limited.

For example, a radioactive particle can either decay or not decay. It can't partially decay. It must be one or the other. You can accurately predict the average time it takes a given type of particle to decay, but you can't in any way determine when a specific, individual particle will decay. Let's say you have 100 similar particles, and you know that 98% of them will decay within time *X*. There is no way to predict in advance which two of the particles will be left after time *X* has expired, but it is a very good bet that there will be exactly two left.

Our double-slit experiment gave us another example. Each particle must pass through slit A or slit B, but you can't predict in advance which way any given particle will go. But you can change the odds by your observation. Normally, 50% of the photons (or other quantum particles) will pass through slit A and 50% will go through slit B. Suppose you place a photon counter at slit A, but not at slit B. Almost all of the photons will go through slit A, and very few, if any, will go through slit B. The interference pattern will not be seen on the screen (or it will be significantly weakened). If you move the photon counter to slit B, you'll just reverse the results. Almost all of the photons shun slit A and pass through slit B.

Now, here's where it really gets weird. What do you suppose happens if you place photon counters at both slits? About 50% of the photons pass through each slit, as before, but there is no (or a very weak) interference pattern on the screen. The particles act just like tiny ping-pong balls, instead of like waves. Somehow, quantum particles ''know'' when they are being observed in a given way and behave accordingly. And once observed, they are locked into that particular pattern.

In effect, with no photon counters, each photon is actually passing through both slits at once. Both paths are equally possible, and both are equally real to the quantum particle. But once you pinpoint the particle to a specific location, all indeterminacy is removed, and the particle is locked into only one specific possibility.

If you're not sure you understand all that, don't worry. No one really understands it. Quantum physicists are the first to admit they don't really understand it. But it's real. Quantum particles actually do work this way.

You might have heard of Schrodinger's cat. This is a famous thought experiment devised by a man named Erwin Schrodinger that illustrates the indeterminacy principle. Suppose you have a cat locked in a sealed box. There is no way to observe the cat until the box is opened. Also in the box is a vial of fast-acting poison gas. A special triggering mechanism is controlled by a single radioactive particle. If it decays, the vial is opened, releasing the gas, and the cat is killed. If the particle doesn't decay, the cat is safe.

Let's say you are using a particle we'll call ''X.'' ''X'' particles have an average half-life of 1 hour. That is, if we start out with 100 ''X'' particles, half of them will decay within one hour, leaving us with 50. Remember, it is impossible to predict when a given particle will decay. Its condition is indeterminate until it is measured.

Now, the cat has been sealed in the box for 1 hour. Because there is just one ''X'' particle, there is a 50% chance that it has decayed and a 50% chance that it has not decayed. So, which has happened? Is the cat alive or dead? In quantum terms, it is both alive *and* dead until the box is opened. Once someone looks inside, the situation is instantly locked into one state or the other, but until the observation is made, all equal possibilities have equal reality. They are simultaneously true and simultaneously untrue in a totally indeterminate state.

As you can see, there is no possibility of any external observation in quantum physics that does not actually affect what is being observed. In a very real sense, the observer creates his own reality, which sounds very, very much like something out of the New Age movement.

There really isn't space in this book to do justice to the complex and confusing ideas of quantum physics, but I hope I have given you some of the flavor. Now I'll leave you with this thought—Is quantum physics a high-tech version of certain New Age ideas and concepts? Could mystics through the ages have somehow stumbled upon some of the peculiar realities of quantum physics long before scientists made the necessary discoveries? If not, why is there so much similarity in the concepts presented in mystic literature through the ages and the recent discoveries of quantum physics? I won't attempt to answer such questions. Frankly, I just don't know, and no one else does either. But it's certainly something to think about, isn't it?

Index

A

air ionization, 79-89
cleaning using negative ions, 82
effects on human beings, 80-82
lightening, 80
negative-ion generator project, 85-88
negative-ion generators, 83-85
airplanes, UFO illusions, 141-142
alpha glasses, 29-33
parts list, 30
schematic diagram, 30
warning for epileptics, 32-33
alpha-wave biofeedback monitor, 51-56
parts list, 54
schematic diagram, 53
alpha waves, 29, 32, 51-52, 55
atomic number, 156
atomic theory, 155-160
atoms, 79, 155-160
audio hypnotizer
improved, 21-23
improved, parts list, 23
improved, schematic diagram, 22
simple, 17-21
simple, parts list, 19
simple, schematic diagram, 18
audio/visual hypnotic aids, 27-28
parts list, 28
schematic diagram, 27
auras, 121-123
different colors, 122
automated ESP tester, 74-77
parts list, 76
schematic diagram, 74

B

beta waves, 29, 51
Bickel, William S., 127
biofeedback, 9, 168 (*see also* hypnosis; meditation)
alpha-wave monitor project, 51-56
body temperature monitor project, 43-47
capacitive monitor project, 47-51
computerized monitor project, 41-43
final thoughts, 56
monitors, 35-56
skin resistance monitor project, 38-41
theory, 35-38
biological functions, 8-9
biorhythm calculation program, 102-108
flowchart, 104
biorhythm clock project, 108-117
parts list, 114
schematic diagrams, 110-113
biorhythms, 9, 91-117
cycles, 9, 92-93, 100-101
definition, 91-94
flaws in reasoning, 97-99
history, 99-102
rhythms in life, 94-97
blue light, 120
body temperature biofeedback monitor, 43-47
parts list, 46
schematic diagram, 45
brain waves, 29, 32, 51, 166-167
alpha, 29, 32, 51-52, 55
beta, 29, 51
delta, 51
theta, 32, 51
Brown, Dr. Barbara, 35

C

capacitance board, 49
capacitive biofeedback monitor, 47-51
parts list, 48
schematic diagram, 48

circadian cycle, 95
clairvoyance, 8, 57-58
  testing, 61, 62
computerized biofeedback monitor, 41-43
  parts list, 42
  schematic diagram, 42
coronas (*see* auras)
critical days, 93
crystal lattice, 159
crystals, 7, 155-172
  ancient history, 164-165
  composition of, 155-160
  energy, 167-168
  healing powers, 169-171
  in electronics, 160-164
  living, 171
  New Age, 164-165
  psionic generators, 172
  resonant frequency, 163
  scientific claims vs. New Age beliefs, 165-171
  symbol, 161
  vibrations, 166-167
cycles
  circadian, 95
  emotional, 92-93, 100
  intellectual, 92-93, 101
  male hormonal, 95
  menstrual, 94, 97-98
  physical, 92-93, 100
  sleep/waking, 95-96

D

delta waves, 51

E

Einstein, Albert, 176
electron, 79, 155, 156
electronic dice, 62-66
  parts list, 64
  schematic diagram, 63
electron ring, 157
emotional cycle, 92-93, 100
epilepsy, 32-33
ESP, 8
  automated tester project, 74-77
  electronic dice project, 62-66
  manual telepathy tester project, 69-73
  random-number generator project, 66-69
  testers, 57-77
  testing, 59-60
  two-choice tester project, 60-62
  types of, 57-59
extrasensory perception (*see* ESP)

F

faith healing, 126
Fliess, Dr. William, 99-100
frequency
  fundamental, 163
  resonant, 163
fundamental frequency, 163

G

galvanic skin response (GSR), 37

H

harmonics, 163
healings, 6-7
  faith, 126
  psychic, 170
  using crystals for, 169-171
hypnosis, 14-17 (*see also* biofeedback)
  deluxe audio/visual hypnotic aid project, 27-28
  dual LED visual hypnotic aid project, 25-27
  improved audio hypnotizer project, 21-23
  origin, 15
  simple audio hypnotizer project, 17-21
  visual hypnotic aid project, 22-25

I

infrared light, 120
intellectual cycle, 92-93, 101
ions, 79-80 (*see also* air ionization)
  negative, 80-89, 156
  positive, 79-82, 156
Ionotron, 81
ion wind, 83

J

jet lag, 96-97

K

Kirlian, Semyon, 120-121

Kirlian photography, 119-136
  auras, 121-123
  diagnostic aura research, 125-127
  experimental circuit project, 128-132
  general information, 120-121
  making the photograph, 132-135
  parts list for project, 130
  phantom leaf effect, 124-125
  possible applications, 123-124
  safety precautions, 135-136
  schematic diagram for project, 129
  skeptics, 127-128

L

lie detector, 37-38
  computerized biofeedback monitor, 41-43
  skin resistance project, 38-41
light, 120
lightening, 80

M

magnetic field, 143-144
magnetic field detector 1 project, 144-149
  parts list, 145
  schematic diagram, 144
magnetic field detector 2 project, 149-153
  parts list, 150
  schematic diagram, 150
male hormonal cycle, 95
mantra, 14
manual telepathy tester, 69-73
  parts list, 72
  schematic diagram, 70-71
meditation (*see also* biofeedback)
  aids, 14
  benefits, 13-14
menstrual cycle, 94, 97-98
Mesmer, F. A., 15
mind reading (*see* telepathy)
Mittelmann, B., 44
molecule, 157
Moss, Dr. Thelma, 126

N

negative ions, 80-89, 156
negative-ion generators, 83-85
  project, 85-88
  project parts list, 87
  project safety considerations, 88-89
  project schematic diagram, 86
neutron, 79, 156
New Age
  beliefs and concepts, 3-9
  vs. scientific theories, 1-3
noble gasses, 158-159
nucleus, 79, 155, 156

O

occult, 4
overtones, 163
oxygen (*see* air ionization)
ozone, 89

P

photography
  Kirlian, 119-136
  nonstandard, 119-120
photon, 176
physical cycle, 92-93, 100
physics
  new/old, 174-175
  quantum, 2, 9-11, 173-183
piezoelectric effect, 161, 168
polycrystalline material, 159
positive ions, 79-82, 156
premenstrual syndrome (PMS), 94
projects
  alpha glasses, 29-33
  audio hypnotizer, 17-23
  automated ESP tester, 74-77
  biofeedback alpha-wave monitor, 51-56
  biofeedback body temperature monitor, 43-47
  biofeedback capacitive monitor, 47-51
  biofeedback computerized monitor, 41-43
  biofeedback skin resistance monitor, 38-41
  biorhythm calculation program, 102-108
  biorhythm clock, 108-117
  deluxe audio/visual hypnotic aid, 27-28

projects *cont.*
- dual LED visual hypnotic aids, 25-27
- electronic dice, 62-66
- Kirlian photography, 128-132
- lie detector, 38-43
- magnetic field detector 1, 144-149
- magnetic field detector 2, 149-153
- manual telepathy tester, 69-73
- meditation, 29-33
- negative-ion generator, 85-88
- random-number generator, 66-69
- two-choice ESP tester, 60-62
- visual hypnotic aids, 22-25

proton, 79, 155, 156
psychic (*see* clairvoyance)

Q

quanta, 11
quantum leap, 180
quantum particles
- behavior, 175-176
- movement, 180-181

quantum physics, 2, 9-11, 173-183
- double slit experiment, 176-180
- indeterminancy principle, 181-183

quartz, 159-160 (*see also* crystals)

R

random-number generator, 66-69
- parts list, 67
- schematic diagram, 67

red light, 120
reincarnation, 5
religion, 4-5
resonant frequency, 163

S

Schrodinger, Erwin, 182
serotonin, 81
shamans, 169-170
sine waves, 92
skin resistance biofeedback monitor, 38-41
- parts list, 39
- schematic diagram, 39

sleep/waking cycle, 95
swamp gas, 141
Swoboda, Dr. Hermann, 99-100

T

telekinesis, 8, 58
telepathy, 8, 57
- manual tester project, 69-73
- testing, 61, 62

Teltscher, Alfred, 101
The Skeptical Inquirer, 127-128
theta waves, 32, 51
two-choice ESP tester, 60-62
- parts list, 61
- schematic diagram, 60

U

UFOs, 7-8, 137-153
- airplane illusions, 141-142
- alien beings, 138-139
- beliefs, 137-143
- detecting, 137-153
- magnetic field detector 1 project, 144-149
- magnetic field detector 2 project, 149-153
- magnetic field effects of, 143-144
- swamp gas, 141
- U.S. Government involvement, 138
- weather balloons, 141

ultraviolet light, 120
unidentified flying objects (*see* UFOs)

V

visual hypnotic aids, 22-25
- dual LED, 25-27
- dual LED, parts list, 26
- dual LED, schematic diagram, 26
- parts list, 24
- schematic diagram, 23

W

Watkins, Arleen J., 127
weather balloons, 141
white light, 120
Wolff, H. G., 44

X

X-rays, 119-120

Y

yogi, 14, 36